STATISTICS: Textbooks and Monographs

A SERIES EDITED BY

D. B. OWEN, Coordinating Editor
Department of Statistics
Southern Methodist University
Dallas, Texas

Volume 1: The Generalized Jackknife Statistic, *H. L. Gray and W. R. Schucany*

Volume 2: Multivariate Analysis, *Anant M. Kshirsagar*

Volume 3: Statistics and Society, *Walter T. Federer*

Volume 4: Multivariate Analysis: A Selected and Abstracted Bibliography, 1957-1972, *Kocherlakota Subrahmaniam and Kathleen Subrahmaniam (out of print)*

Volume 5: Design of Experiments: A Realistic Approach, *Virgil L. Anderson and Robert A. McLean*

Volume 6: Statistical and Mathematical Aspects of Pollution Problems, *John W. Pratt*

Volume 7: Introduction to Probability and Statistics (in two parts) Part I: Probability; Part II: Statistics, *Narayan C. Giri*

Volume 8: Statistical Theory of the Analysis of Experimental Designs, *J. Ogawa*

Volume 9: Statistical Techniques in Simulation (in two parts), *Jack P. C. Kleijnen*

Volume 10: Data Quality Control and Editing, *Joseph I. Naus (out of print)*

Volume 11: Cost of Living Index Numbers: Practice, Precision, and Theory, *Kali S. Banerjee*

Volume 12: Weighing Designs: For Chemistry, Medicine, Economics, Operations Research, Statistics, *Kali S. Banerjee*

Volume 13: The Search for Oil: Some Statistical Methods and Techniques, *edited by D. B. Owen*

Volume 14: Sample Size Choice: Charts for Experiments with Linear Models, *Robert E. Odeh and Martin Fox*

Volume 15: Statistical Methods for Engineers and Scientists, *Robert M. Bethea, Benjamin S. Duran, and Thomas L. Boullion*

Volume 16: Statistical Quality Control Methods, *Irving W. Burr*

Volume 17: On the History of Statistics and Probability, *edited by D. B. Owen*

Volume 18: Econometrics, *Peter Schmidt*

Volume 19: Sufficient Statistics: Selected Contributions, *Vasant S. Huzurbazar (edited by Anant M. Kshirsagar)*

Volume 20: Handbook of Statistical Distributions, *Jagdish K. Patel, C. H. Kapadia, and D. B. Owen*

Volume 21: Case Studies in Sample Design, *A. C. Rosander*

Volume 22: Pocket Book of Statistical Tables, *compiled by R. E. Odeh, D. B. Owen, Z. W. Birnbaum, and L. Fisher*

Volume 23: The Information in Contingency Tables, *D. V. Gokhale and Solomon Kullback*

Volume 24: Statistical Analysis of Reliability and Life-Testing Models: Theory and Methods, *Lee J. Bain*

Volume 25: Elementary Statistical Quality Control, *Irving W. Burr*

Volume 26: An Introduction to Probability and Statistics Using BASIC, *Richard A. Groeneveld*

Volume 27: Basic Applied Statistics, *B. L. Raktoe and J. J. Hubert*

Volume 28: A Primer in Probability, *Kathleen Subrahmaniam*

Volume 29: Random Processes: A First Look, *R. Syski*

Volume 30: Regression Methods: A Tool for Data Analysis, *Rudolf J. Freund and Paul D. Minton*

Volume 31: Randomization Tests, *Eugene S. Edgington*

Volume 32: Tables for Normal Tolerance Limits, Sampling Plans, and Screening, *Robert E. Odeh and D. B. Owen*

Volume 33: Statistical Computing, *William J. Kennedy, Jr. and James E. Gentle*

Volume 34: Regression Analysis and Its Application: A Data-Oriented Approach, *Richard F. Gunst and Robert L. Mason*

Volume 35: Scientific Strategies to Save Your Life, *I. D. J. Bross*

Volume 36: Statistics in the Pharmaceutical Industry, *C. R. Buncher and J.-Y. Tsay*

Volume 37: Sampling from a Finite Population, *J. Hájek*

OTHER VOLUMES IN PREPARATION

SAMPLING FROM A FINITE POPULATION

Jaroslav Hájek (1926-1974)

SAMPLING FROM A FINITE POPULATION

JAROSLAV HÁJEK
Charles University
Prague, Czechoslovakia

Edited by Václav Dupač
Charles University
Prague, Czechoslovakia

MARCEL DEKKER, INC. New York and Basel

Library of Congress Cataloging in Publication Data

Hajek, Jaroslav, [date]
Sampling from a finite population.

(Statistics, textbooks and monographs ; v. 37)
Bibliography: p.
Includes index.
1. Sampling (Statistics) I. Dupac, Vaclav.
II. Title. III. Series.
QA276.6.H334 519.5 81-7835
ISBN 0-8247-1291-9 AACR2

MARCEL DEKKER, INC.
270 Madison Avenue, New York, New York 10016

Current printing (last digit):
10 9 8 7 6 5 4 3 2

PRINTED IN THE UNITED STATES OF AMERICA

6/17/85 pm

FOREWORD

The present monograph by the late professor Jaroslav Hájek (1926-1974) deals primarily with theory of sampling from a finite population, which plays an important and increasing role in statistical theory as well as in applications. In the majority of practical problems, finite population sampling underlies the statistical methodology, and the study of the various characteristics of the population in some desirable or optimal way, along with prescribed margins of error of the estimators, constitutes a basic domain of statistical planning and inference. The present monograph is devoted to these fundamental aspects of finite population sampling.

The late Professor Hájek earned the highest international reputation for his outstanding contributions to the theory of probability, parametric estimation, nonparametrics, statistical inference in stochastic processes, probabilistic sampling, as well as some other applied areas. Indeed, he was one of the most versatile statisticians, who not only contributed generously to various fields but also was able to provide natural bridges between them. In the sixties, he became deeply interested in the development of the theory of probabilistic sampling and made a valuable contribution to this area through a set of invaluable papers in the Annals of Mathematical Statistics and some other Czechoslovak journals. His last paper on sampling with varying probabilities without replacements in 1973 depicts his interest in this field until his untimely death in 1974.

In the early seventies, during his visit to the United States, he became interested in writing a monograph on finite population sampling and pursued the project almost to completion, in spite of his serious kidney problems. He literally worked on this project and guided research until the last days of his life. After his premature death, his colleagues and former students at the Charles University, Prague, Czechoslovakia, put together the final revisions of this monograph.

The publication of this monograph naturally was planned a few years ago, but was delayed, not only because of Professor Hájek's death, but also because of other factors, though the ideas contained in it continued to stimulate interest not only among colleagues at Prague but also in the United States as well as other parts of Europe. Mrs. Betty Hájková's sincere effort in putting together the last work of her late husband as a monograph has met with a good response (and profound respect).

I have no doubt in my mind that the monograph will be very useful to a variety of statisticians desiring to understand the basic principles of finite population sampling and to apply them fruitfully in practical situations. Like the other books of Hájek, the current one reveals his lucid style and presentation at an intermediate level of mathematical sophistication. I also believe that this monograph will stimulate further research in this area of great theoretical and practical interest .

P. K. Sen
Chapel Hill, North Carolina

CONTENTS

SAMPLING FROM A FINITE POPULATION

Part I

BASIC IDEAS AND CONCEPTS

1

POPULATION

A collection of certain units is called a population if the units are not interesting by themselves but only as contributors to statistical properties of the whole. Thus, regarding an aggregate as a population, we do not ask which units have this or that attribute but only how many of them possess it.

Example 1.1 A book is not a mere population of words if we investigate its esthetic or informational contents. However, it becomes a population of words if we explore the frequency of pronouns or some other statistical property. Similarly a collection of workers in a factory is a population if the average age is considered, but is something more if the payroll is compiled.

Usually one is told that the units making up a population must be to a considerable degree similar. The positive meaning of this requirement is that there should be variables (characteristics) of interest that are applicable to all units making up the given population.

Given a population S and a variable y, the most trivial statistical issue is to establish the total

$$Y = y_1 + \cdots + y_N \tag{1.1}$$

where y_i denotes the value of y possessed by the ith unit. However, if the information about the values y_i is not easily accessible and if the number of units N is very large, the practical completion of this task may be rather expensive and time-absorbing. Then we may exploit the fact that Y is a statistical characteristic and estimate it from a sample. This book will be devoted to discussion of various sampling and estimating strategies.

The units making up the population S may be any elements worth studying--persons, families, farms, account items, temperature readings, and so on--and their nature will be irrelevant for theoretical considerations. We shall assume that the units are identifiable by certain labels (tags, names, addresses) and that we have available a frame (list, map) showing how to reach any unit given its label. The simplest labels are the numbers 1, 2, ..., N, and as a matter of fact we shall identify the units with their ordinal numbers by putting $S = \{1,2,\ldots,N\}$.

Also the variable y may denote anything of interest: income, voting preference, number of children, temperature, and so on. In particular, y may indicate the presence or absence of a certain attribute A, that is, $y_i = 1$ if the unit i has the attribute, and $y_i = 0$ if it does not. In that case y will be called the *indicator* of the attribute A. Then, obviously

$$Y = y_1 + \cdots + y_N = N_A \tag{1.2}$$

where N_A is the number of units possessing the property A. We shall call N_A the *frequency* of A. A characteristic that classifies the units according to k mutually exclusive attributes (k different religions, for example) may be expressed by k corresponding indicators.

Example 1.2 In a population consisting of 180 men the circumference of their heads was measured. The results are presented in Table 1.1, where we can read that the total equals

$$Y = 10{,}179 \text{ cm}$$

Example 1.3 Table 1.2 shows frequencies of 14 attributes characterizing dental state and health in the left lower quadrant of the mouth of 786 twelve-year-old girls in Czechoslovakia in the year 1955.

TABLE 1.1 Circumference of head in men

Circumference of head (cm)	Frequency	Product
52	1	52
53	3	159
54	11	594
55	22	1,210
56	56	3,136
57	35	1,995
58	41	2,378
59	6	354
60	4	240
61	1	61
Totals:	180	10,179

Dividing the total Y by the size N of the population, we obtain the *average*

$$\bar{Y} = \frac{Y}{N} = \frac{1}{N}(y_1 + \cdots + y_N) \tag{1.3}$$

If y is the indicator of an attribute, then the average is identical with the *relative frequency* of the attribute in the population. We shall denote the relative frequency by P:

$$P = \frac{N_A}{N} \tag{1.4}$$

TABLE 1.2 Dental state and health in 12-year-old girls (N = 786)

Tooth number	1	2	3	4	5	6	7
Frequency of girls with the tooth cut through	782	785	779	759	686	785	702
Frequency of girls with the tooth cut through and healthy	760	770	777	745	617	109	379

Example 1.4 The average circumference of the head in men computed from the data of Table 1.1 equals

$$\bar{Y} = \frac{10,179}{180} = 56.6 \text{ cm}$$

The relative frequency of girls that have the tooth number 5 in the left lower quadrant cut through is, according to Table 1.2,

$$P = \frac{686}{786} = 0.873 = 87.3\%$$

The average is a special case of a ratio of two totals:

$$R = \frac{Y}{X} \tag{1.5}$$

where $X = x_1 + \cdots + x_N$ for some another variable x. Ratio characteristics are capable of expressing indices of growth, intensities, proportions--generally, the relative magnitude of y in terms of x.

Example 1.5 An important anthropometric characteristic is the ratio of the width and length of the face. The data given in Table 1.3 provide the following value for it:

$$R = \frac{\text{total of face lengths}}{\text{total of face widths}} = \frac{25,630}{21,936} = 1.17$$

Example 1.6 A natural way to judge the attribute "the tooth is cut through and healthy" is in terms of the attribute "the tooth is cut through." Table 1.4 provides values for the corresponding ratio characteristics for teeth numbers 1, 2, ..., 7.

The variability of y is usually measured by the *variance*

$$\sigma_y^2 = \frac{1}{N} \sum_{i=1}^{N} (y_i - \bar{Y})^2 \tag{1.6}$$

and, for nonnegative variables, the relative variability by the coefficient of variation

$$V_y = \frac{\sigma_y}{\bar{Y}} \tag{1.7}$$

TABLE 1.3 Men classified according to the width and length of the face (N = 179)

Width (mm)	Length (mm)											Conditional totals	
	131	134	137	140	143	146	149	152	155	158	161	Frequency	Length
108		1	1									2	271
111			1	1	1		1					4	569
114		1		2	2	2	2					9	1,290
117		3	4	9	9	9	3				1	38	5,419
120		1	6	4	9	6	2					28	3,977
123	1	1	3	6	11	3	3	1				29	4,126
126			2	3	3	6	10	2			1	27	3,954
129		2	6	6	8	2	4		1	1		30	4,275
132				1	3	1	1	1		1		8	1,174
135						1	1					2	295
138							1					1	149
141	1											1	131
Conditional totals:												Grand total:	
Frequency:	2	9	23	32	46	30	28	4	1	2	2	179	25,630
Width :	264	1,074	2,802	3,894	5,631	3,651	3,480	507	129	261	243	21,936	

TABLE 1.4 Percentage of healthy teeth among cut-through ones for 12-year-old girls and teeth Nos. 1, 2, ..., 7

Tooth number	Percentage
1	760/782 = 97.2%
2	770/785 = 98.1
3	777/779 = 99.7
4	745/759 = 98.2
5	617/686 = 89.9
6	109/785 = 13.9
7	379/702 = 54.0

Note: A tooth is qualified as healthy if free of cavities as well as of fillings.

The square root of σ_y^2, i.e., σ_y, is called the *standard deviation.* The joint variability of y and x may be expressed by the *covariance*

$$\sigma_{yx} = \frac{1}{N} \sum_{i=1}^{N} (y_i - \bar{Y})(x_i - \bar{X}) \tag{1.8}$$

Dividing σ_{yx} by $\sigma_y \sigma_x$, we obtain the *correlation coefficient*

$$\rho = \frac{\sigma_{yx}}{\sigma_y \sigma_x} \tag{1.9}$$

which measures the extent of linear dependence of y on x.

Dividing by σ_x^2, we obtain the *coefficient of regression* of y on x:

$$\beta_{yx} = \frac{\sigma_{yx}}{\sigma_x^2} \tag{1.10}$$

We can see that the *regression line*

$$y = \bar{Y} + \beta_{yx}(x - \bar{X}) \tag{1.11}$$

passes through the origin, i.e., is equivalent to $y = \beta_{yx}x$, iff

$$\beta_{yx} = \frac{Y}{X} \tag{1.12}$$

If y is the indicator of an attribute A and $Y = N_A$, then the variance and the variation coefficient equal

$$\sigma_A^2 = \frac{N_A(N - N_A)}{N \cdot N} = P_A(1 - P_A) \tag{1.13}$$

$$V_A = \sqrt{\frac{N(N - N_A)}{N_A N}} = \sqrt{\frac{1}{P_A} - 1} \tag{1.14}$$

where $P_A = N_A/N$.

If y and x indicate attributes A and B, respectively, then their correlation coefficient equals

$$\rho_{AB} = \frac{P_{AB} - P_A P_B}{\sqrt{P_A(1 - P_A)P_B(1 - P_B)}} \tag{1.15}$$

where P_A, P_B, P_{AB} are relative frequencies of attributes A, B and of joint occurrence of A and B.

PROBLEMS AND NOTES

1.1 Compute the regression coefficient β_{yx} from Table 1.3 if y denotes the length and x denotes the width of the face. Show that it greatly differs from $R = Y/X$. Draw the regression line of y on x and observe that it passes far from the origin.

1.2 Compute β_{yx} and $R = Y/X$ from the data of Table 1.5. Note that the regression nearly passes through the origin.

1.3 Let y and x indicate two attributes A and B such that A occurs only on those units where B does. Then the regression of y on x passes through the origin if $N_A > 0$.

1.4 Let y and x indicate two attributes that exclude each other. Then the correlation coefficient of y and x is negative. When does it equal -1?

TABLE 1.5 Areas under wheat and yields of trade wheat in 37 Bohemian counties, 1928

County	Area (10 ha) x_i	Yield (10,000 kg) y_i
1	210	477
2	130	316
3	377	883
4	38	82
5	282	574
6	127	327
7	137	347
8	116	282
9	145	369
10	116	280
11	75	131
12	392	941
13	191	453
14	270	620
15	274	746
16	188	385
17	297	740
18	219	494
19	203	537
20	286	728
21	269	684
22	300	754
23	372	1,021
24	188	402
25	221	509
26	291	769
27	205	482
28	258	567
29	269	625
30	219	489
31	315	660
32	353	852
33	384	905
34	176	344
35	250	605
36	206	462
37	205	358

1.5 Let $(y_i \le y_j) \Leftrightarrow (x_i \le x_j)$ for all pairs $1 \le i \le j \le N$. Then $\rho_{yx} \ge 0$.

More generally, $\rho_{yx} \ge 0$ also holds if $(\bar{Y}_x \le \bar{Y}_{x'}) \Leftrightarrow (x \le x')$, where $\bar{Y}_x$ denotes the partial average of y in the subpopulation where $x_i = x$.

1.6 *Definitions* (a) A *weighted* average, variance, and coefficient of variation of some numbers $y_1, \ldots, y_N$ with weights $w_1, \ldots, w_N$, where $w_i \ge 0$, $1 \le i \le N$, $\Sigma_{i=1}^N w_i > 0$, is defined as follows, respectively:

$$\tilde{Y} = \frac{w_1 y_1 + \cdots + w_N y_N}{w_1 + \cdots + w_N} \tag{1.16}$$

$$\tilde{\sigma}^2 = \frac{\sum_{i=1}^N (y_i - \tilde{Y})^2 w_i}{\sum_{i=1}^N w_i} \tag{1.17}$$

$$\tilde{V} = \frac{\tilde{\sigma}}{\tilde{Y}} \tag{1.18}$$

The weighted covariance and correlation coefficient are defined similarly.

(b) The square of the variation coefficient, $V_y^2 = \sigma_y^2/\bar{Y}^2$, is called a *relative variance* (or rel-variance, for short). Similarly, we can introduce the *relative covariance* (rel-covariance) $V_{yx} = \sigma_{yx}/\bar{Y}\bar{X}$.

1.7 If the population is partitioned into L disjoint parts (strata) of respective sizes $N_1, \ldots, N_L$ and partial averages equaling $\bar{Y}_1, \ldots, \bar{Y}_L$, then

$$\bar{Y} = \frac{\bar{Y}_1 N_1 + \cdots + \bar{Y}_L N_L}{N_1 + \cdots + N_L} \tag{1.19}$$

Consequently, the grand average is a weighted average of the strata averages with weights $N_1, \ldots, N_L$.

1.8 *The Cauchy inequality* If $u_1, \ldots, u_L$ and $v_1, \ldots, v_L$ are arbitrary nonnegative numbers, then

$$(u_1 + \cdots + u_L)(v_1 + \cdots + v_L) \geq (\sqrt{u_1 v_1} + \cdots + \sqrt{u_L v_L})^2 \tag{1.20}$$

More precisely, we have

$$\sum_{h=1}^{L} u_h \sum_{h=1}^{L} v_h = \left(\sum_{h=1}^{L} \sqrt{u_h v_h}\right)^2 (1 + V_*^2) \tag{1.21}$$

where V_*^2 is the weighted rel-variance of the numbers $\sqrt{u_1/v_1}$, ..., $\sqrt{u_h/v_h}$ with weights v_1, ..., v_L.

1.9 *The Jensen inequality* If f is a convex function on (a,b) [for example, if $f''(y) \geq 0$ for all $y \in (a,b)$] and if all values y_i lie in (a,b), then

$$\frac{1}{N} \sum_{i=1}^{N} f(y_i) \geq f(\bar{Y}) \tag{1.22}$$

Apply it to $f(y) = y^2$, $f(y) = -\ln y$, and $f(y) = y \ln y$.

1.10 Let a + by denote the variable that takes the value $a + by_i$ when y takes y_i. Then

$$\sigma_{a+by}^2 = b^2 \sigma_y^2 \tag{1.23}$$

$$\sigma_{a+by,c+dx} = bd\sigma_{yx} \tag{1.24}$$

$$\rho_{a+by,c+dx} = \rho_{yx} \qquad \text{if } bd > 0 \tag{1.25}$$

$$V_{by} = V_y \qquad \text{if } b > 0 \tag{1.26}$$

1.11 If f is a function that is smooth enough in the vicinity of $\bar{Y}$, then

$$\overline{f(Y)} \doteq f(\bar{Y}) \qquad \sigma_{f(y)} \doteq |f'(\bar{Y})| \sigma_y \tag{1.27}$$

Applying this result, show that

$$V_{y^\alpha} \doteq \alpha V_y \qquad \text{for } y_i \geq 0 \qquad 1 \leq i \leq N \tag{1.28}$$

Further show that

$$\sigma_{\log y} \doteq V_y \qquad y_i > 0 \qquad 1 \leq i \leq N \tag{1.29}$$

1.12 Prove the following identities:

$$N\sigma_y^2 = \sum_{i=1}^{N} (y_i - \bar{Y})y_i = \sum_{i=1}^{N} (y_i^2 - \bar{Y}^2) \tag{1.30}$$

$$N\sigma_y^2 = \sum_{i=1}^{N} [(y_i - a)^2 - (\bar{Y} - a)^2] \tag{1.31}$$

$$N\sigma_y^2 = \frac{1}{2N} \sum_{i=1}^{N} \sum_{j=1}^{N} (y_i - y_j)^2 \tag{1.32}$$

$$N\sigma_y^2 = \sum_{h=1}^{L} \sum_{i \in S_h} (y_i - \bar{Y}_h)^2 + \sum_{h=1}^{L} (\bar{Y}_h - \bar{Y})^2 N_h \tag{1.33}$$

where a is an arbitrary number and Eq. (1.33) refers to a partition into L strata S_h described in Problem 1.7.

1.13 Prove the following identities:

$$N\sigma_{yx} = \sum_{i=1}^{N} (y_i - \bar{Y})x_i = \sum_{i=1}^{N} (y_i x_i - \bar{Y}\bar{X}) \tag{1.34}$$

$$N\sigma_{yx} = \sum_{i=1}^{N} [(y_i - b)(x_i - a) - (\bar{Y} - b)(\bar{X} - a)] \tag{1.35}$$

$$N\sigma_{yx} = \frac{1}{2N} \sum_{i=1}^{N} \sum_{j=1}^{N} (y_i - y_j)(x_i - x_j) \tag{1.36}$$

$$N\sigma_{yx} = \sum_{h=1}^{L} \sum_{i \in S_h} (y_i - \bar{Y}_h)(x_i - \bar{X}_h) - \sum_{h=1}^{L} (\bar{Y}_h - \bar{Y})(\bar{X}_h - \bar{X})N_h \tag{1.37}$$

where a and b are arbitrary numbers and Eq. (1.37) refers again to a partition into strata.

2

BAYESIAN APPROACH: PROBABILITY SPECULATION

Valid inference from the sample to the population must have a probabilistic basis. We have to find some link connecting the observed values with the unknown ones. One approach to this aim consists in speculating on $y_1, \ldots, y_N$ as *random variables*, whose distribution is either completely known, or involves few parameters, or is known to belong to some simple nonparametric family. If our prior information is rich enough, any reasonably large sample of the population may serve as a safe basis for the estimation of the population total Y. The selection of the sample may then be *purposive*, dictated only by economy and required precision. No randomness needs to be introduced into the selection procedure. However, the catch is in the unfortunate circumstance that sufficiently informative speculations on $y_1, y_2, \ldots, y_N$ are rarely realistic and may be easily misleading for sample sizes used in sample surveys. Thus probability speculation, though very fruitful, provides an insufficient basis for statistical inference and it must be complemented by *probability sampling*, discussed in the next chapter. Probability sampling preserves the advantages of probability speculation while removing its possible inadequacies.

Probability speculation is equivalent to regarding our population $y_1, \ldots, y_N$ as selected from a set of populations that were also possible under given circumstances. This set of populations is often called a *superpopulation* and the family of distributions of $y_1, \ldots, y_N$ that are a priori admissible is called the *superpopulation model*.

Before giving some examples let us denote the expectation and variance connected with the distribution of $y_1, \ldots, y_N$ by M and D, respectively. These terms will have the standard meaning. For example, if z is a random variable assuming values $c_1, \ldots, c_k$ with probabilities $p_1, \ldots, p_k$, then

$$Mz = c_1p_1 + \cdots + c_kp_k \tag{2.1}$$

If z is of continuous type with density f(c), then

$$Mz = \int_{-\infty}^{\infty} cf(c)\, dc \tag{2.2}$$

Also the terms *covariance*, *independence*, etc. will be used in the usual sense. The normal distribution with expectation a and variance b^2 will be denoted as $N(a,b^2)$. It is known to possess the density

$$f(u) = \frac{1}{b\sqrt{2\pi}} \exp\left[-\frac{(u-a)^2}{2b^2}\right] \qquad -\infty < u < \infty \tag{2.3}$$

There is a close connection between the concepts of the descriptive statistic discussed in the previous section and the probabilistic concepts. For example, the expectation as given in (2.1) may be considered as a weighted average of $c_1, \ldots, c_k$ with weights equaling the probabilities $p_1, \ldots, p_k$.

Example 2.1 Assume that $y_1, \ldots, y_N$ are independent random variables all having the same normal distribution $N(\mu,b^2)$, where μ and b^2 are unknown parameters. Further assume that the sample consists of the first n units so that the problem is to estimate $Y = y_1 + \cdots + y_N$ on the basis of $(y_1,\ldots,y_n)$. Since $y_1 + \cdots + y_n$ is known, we may equivalently estimate $y_{n+1} + \cdots + y_N$. Denoting the respective estimate as $t(y_1,\ldots,y_n)$, we can easily see that

$$\begin{aligned} &M[y_{n+1} + \cdots + y_N - t(y_1,\ldots,y_n)]^2 \\ &= M[(N-n)\mu - t(y_1,\ldots,y_n)]^2 + (N-n)b^2 \end{aligned} \tag{2.4}$$

Thus the problem reduces to the estimation of μ--the expectation in a normal sample. This is a standard problem yielding the estimate $\bar{y} = (1/n)(y_1 + \cdots + y_n)$ for μ and the estimate $(N - n)\bar{y}$ for $(N - n)\mu$. Thus

$$t(y_1,\ldots,y_n) = (N - n)\bar{y} \tag{2.5}$$

Consequently $y_{n+1} + \cdots + y_N$ will be estimated by $(N - n)\bar{y}$ and Y by $N\bar{y}$. Denoting the estimate of Y by $\hat{Y}$, we can write

$$\hat{Y} = \frac{N}{n}(y_1 + \cdots + y_n) \tag{2.6}$$

The mean square error equals

$$M(\hat{Y} - Y)^2 = \frac{b^2 N(N - n)}{n} \tag{2.7}$$

as is easily seen from (2.4) and the fact that

$$M[(N - n)\mu - (N - n)\bar{y}]^2 = \frac{b^2(N - n)^2}{n} \tag{2.8}$$

Utilizing the standard estimate for b^2, we obtain the following estimate $v(\omega)$ for $M(\hat{Y} - Y)^2$:

$$v(\omega) = N(N - n)\frac{\sum_{i=1}^{n}(y_i - \bar{y})^2}{n(n - 1)} \tag{2.9}$$

where $\omega = (y_1,\ldots,y_n)$. The Student ratio

$$t = \frac{\hat{Y} - Y}{\sqrt{v(\omega)}} \tag{2.10}$$

has the Student distribution with $n - 1$ degrees of freedom for all $n < N$.

Example 2.2 (Nonparametric model) We shall again assume that the y_i terms are independent with common distribution F. This time, however, the distribution will be arbitrary, provided it has a positive and finite variance. Then it may be shown that the Student ratio (2.10) where $\hat{Y}$ and $v(\omega)$ are defined again by (2.6) and (2.9) respectively, has for $n \to \infty$, $(N - n) \to \infty$, an asymptotically normal distribution N(0,1) no matter which F holds. Thus inferences based on t are robust, in a certain sense.

Example 2.3 Consider a population of trees and let y denote the volume of timber. For this population the model of Example 2.1 is hardly applicable, since there may be places in the forest where trees grow larger than elsewhere, i.e., the forest is not *well mixed*. However, if we are supplied for every tree an eye estimate of the volume of its timber, say x_i, we may assume that the y_i terms are independent random variables that have the normal distributions $N(ax_i, b^2x_i)$, where $a,b > 0$ are unknown parameters.

The same technique as in Example 2.1 will yield that the best estimate for Y will be

$$\hat{Y} = X \frac{y_1 + \cdots + y_n}{x_1 + \cdots + x_n} = X \frac{y}{x} \tag{2.11}$$

if $X = x_1 + \cdots + x_N$, $y = y_1 + \cdots + y_n$, $x = x_1 + \cdots + x_n$ and the trees in the sample are denoted by numbers 1, 2, ..., n. The estimate (2.11) is the so-called ratio estimate. The mean square error will equal

$$M(\hat{Y} - Y)^2 = \frac{b^2X(X - x)}{x} \tag{2.12}$$

Substituting the standard estimate for b^2, namely

$$\hat{b}^2 = \frac{1}{n - 1} \sum_{i=1}^{n} \frac{[y_i - (y/x)x_i]^2}{x_i} \tag{2.13}$$

we obtain the following estimate for $M(\hat{Y} - Y)^2$:

$$v(\omega) = X(X - x) \sum_{i=1}^{n} \frac{[y_i - (y/x)x_i]^2}{x_i x(n-1)} \tag{2.14}$$

Again the Student ratio (2.10) will have the Student distribution with n - 1 degrees of freedom.

The mean square error in (2.12) depends on the sample through the sample total x of the x values. It decreases with increasing x so that the best sample of given size will consist of the n *largest* units. One can easily see that such a sampling strategy may be fatal if our model is not completely correct--for example, if the regression of y on x is not linear and passing through the origin but quadratic. Any mistake in the model will be the more harmful the larger the sample size. This illustrates well the pitfalls of the probability speculation approach.

PROBLEMS AND NOTES

2.1 Assume that the y_i terms are independent and distributed $N(ax_i, b^2 x_i^\lambda)$, $1 \le i \le N$, where $x_1, \ldots, x_N$ and λ are known nonnegative numbers and $a, b^2 > 0$ are unknown parameters. Justify the estimates

$$\hat{Y} = y + \hat{a}(X - x) \tag{2.15}$$

$$v(\omega) = \hat{M}(\hat{Y} - Y)^2 = \left[\sum_{i=n+1}^{N} x_i^\lambda + \frac{(X - x)^2}{\sum_{i=1}^{n} x_i^{2-\lambda}} \right] \sum_{i=1}^{n} \frac{(y_i - \hat{a}x_i)^2}{(n-1)x_i^\lambda} \tag{2.16}$$

where $y = y_1 + \cdots + y_n$, $x = x_1 + \cdots + x_n$ and

$$\hat{a} = \frac{\sum_{i=1}^{n} y_i x_i^{1-\lambda}}{\sum_{i=1}^{n} x_i^{2-\lambda}} \tag{2.17}$$

2.2 Assume that the y_i terms are independent and distributed $N(a_0 + \Sigma_{j=1}^{m} a_j x_{ji}, b^2)$ where $a_0, a_1, \ldots, a_m$ and b^2 are unknown parameters and x_{ji}, $1 \le j \le m$, $1 \le i \le N$ are known numbers. Show that for any unbiased estimate $\hat{Y}$ that is linear in $y_1, \ldots, y_n$ and any sample of size n we have

$$M(\hat{Y} - Y)^2 \ge \frac{b^2 N(N - n)}{n} \tag{2.18}$$

and that equality occurs for

$$\hat{Y} = \frac{N}{n} y$$

and for all samples such that the sample average equals the population average for all concomitant variables "x_1", ..., "x_m".

3

ROBUST APPROACH: PROBABILITY SAMPLING

Simple statistical models rarely describe properly the generation of large-scale statistical populations. These populations are simply too large and consequently they reveal inadequacies of simple models like a magnifying glass. The most dubious point of simple models usually is the assumption of independence. If we try to take care of possible dependencies by introducing stochastic processes, we are likely to sink into complicated estimation problems and still can never be sure that all important features were covered.

Of course, the above argument is affected not only by population size but also by sample size. If we select only a few units, the inference will be invalidated by improper modeling far less than in huge samples. Furthermore, as a rule, the linear regression residuals $y_i - \Sigma_{j=1}^{m} x_{ji}\beta_j$ behave usually more regularly from the stochastic point of view than the original values y_i, and the more regressors x_j used the better. Thus the limits within which we can rely on probability speculation if sampling a population depend on

1. How well the population is mixed
2. How large the sample is
3. How many regressors are used

The attitude taken in this book is that in sample surveys these conditions are met in a satisfactory combination very rarely and, therefore, we should not make use of purposive sampling based solely on probability speculation.

A firm basis for statistical inference in sample surveys may be achieved by *probability sampling*, which introduces a rigorous element of randomness into the sampling procedure. However, there is not just one possible probability sampling, but a rich family of possible probability sampling methods. The choice of the most suitable sampling method in a given situation is the proper sphere for utilizing probability speculation described in the previous chapter. A suitable probability sampling design gives us means of exploiting the advantages of probability speculation and simultaneously remaining immune against its fallacies.

PROBABILITY SAMPLING The formal definition of probability sampling may be given in various ways, but the following is particularly simple:

The sample s will be defined as an arbitrary subset of the population S. As is well known, there are 2^N subsets of $\{1,2,\ldots,N\}$ if we count the empty subset and also the whole set $s = S$. The ordering of the units $i_1, \ldots, i_n$ in $s = \{i_1,\ldots,i_N\}$ is considered as irrelevant, and to achieve uniqueness in notation we shall assume that $i_1 < \cdots < i_n$. We shall consistently use the braces $\{\cdot\}$ to distinguish sets from ordered sets, denoted by $(\cdot)$.

The *sampling experiment* will yield one sample s in accordance with a certain probability distribution $P(s)$, $s \subset S$. Probabilities $P(s)$ are nonnegative numbers satisfying

$$\Sigma_s \, P(s) = 1 \tag{3.1}$$

where the summation Σ_s extends over all 2^N samples. The probability distribution $\{P(s),\ s \subset S\}$ will be also called the *sampling design* (plan, procedure, method). Sampling according to preassigned probabilities $P(s)$ will be called *probability sampling*.

In probability sampling the collection of all 2^N samples s is a probability space and the most important random variables defined on it are indicators of inclusion defined below.

INCLUSION INDICATORS Let $s \ni i$ denote the event that s contains the unit i. Put

$$I_i(s) = \begin{cases} 1 & \text{if } s \ni i \\ 0 & \text{otherwise} \end{cases} \qquad 1 \le i \le N \tag{3.2}$$

and call I_i the indicator of inclusion of the unit i into the sample.

INCLUSION PROBABILITIES The most important characteristics of a probability sampling are the inclusion probabilities. Put

$$\pi_i = \sum_{s \ni i} P(s) \qquad 1 \le i \le N \tag{3.3}$$

$$\pi_{ij} = \sum_{s \ni i,j} P(s) \qquad 1 \le i \ne j \le N \tag{3.4}$$

where the sum $\Sigma_{s \ni i}$ extends over all s containing i and the sum $\Sigma_{s \ni i,j}$ extends over all s containing *both* i and j. We shall call π_i and π_{ij} the first- and second-order probabilities of inclusion, respectively, since $\pi_i = P(s \ni i)$, $\pi_{ij} = P(s \ni i, s \ni j)$.

The following theorem shows that the first and second moments of inclusion indicators may be expressed through inclusion probabilities. We shall denote by $E(\cdot)$, $Var(\cdot)$, and $Cov(\cdot,\cdot)$ the expectation, variance, and covariance, respectively, connected with probability sampling.

THEOREM 3.1 We have

$$EI_i = \pi_i \qquad 1 \le i \le N \tag{3.5}$$

$$\text{Var } I_i = \pi_i(1 - \pi_i) \qquad 1 \le i \le N \tag{3.6}$$

$$\text{Cov}(I_i, I_j) = \pi_{ij} - \pi_i\pi_j \qquad 1 \le i \ne j \le N \tag{3.7}$$

Proof. We have

$$EI_i = \Sigma_s I_i(s)P(s) = \sum_{s \ni i} P(s) = \pi_i$$

$$\text{Var } I_i = EI_i^2 - (EI_i)^2 = EI_i - (EI_i)^2 = \pi_i - \pi_i^2 = \pi_i(1 - \pi_i)$$

$$\text{Cov}(I_i, I_j) = EI_iI_j - EI_iEI_j = \Sigma_s I_i(s)I_j(s)P(s) - \pi_i\pi_j$$

$$= \sum_{s \ni i,j} P(s) - \pi_i\pi_j = \pi_{ij} - \pi_i\pi_j \qquad \text{Q.E.D.}$$

The sample size s may obviously be expressed as the sum of indicators:

$$\text{Size } s = \sum_{i=1}^{N} I_i(s) \tag{3.8}$$

Hence, the expected sample size is the sum of inclusion probabilities, i.e.,

$$E(\text{size } s) = \sum_{i=1}^{N} \pi_i \tag{3.9}$$

For sampling designs where the sample size is fixed, i.e.,

$$P(\text{size } s = n) = 1 \tag{3.10}$$

we have the following useful relations:

THEOREM 3.2 Under (3.10) we have

$$n = \sum_{i=1}^{N} \pi_i \tag{3.11}$$

$$\sum_{j\neq i}^{N} (\pi_i\pi_j - \pi_{ij}) = \pi_i(1 - \pi_i) \qquad 1 \le i \le N \tag{3.12}$$

Proof. From (3.10) we get $\Sigma_{j\neq i}^{N} I_j = n - I_i$. Hence, making use of (3.7),

$$\pi_i(1 - \pi_i) = \text{Var}(I_i) = \text{Cov}(I_i, I_i) = -\text{Cov}(I_i, n - I_i)$$

$$= -\sum_{j\neq i}^{N} \text{Cov}(I_i, I_j) = \sum_{j\neq i}^{N} (\pi_i\pi_j - \pi_{ij})$$

which proves (3.12). Equation (3.11) follows immediately from Eq. (3.9).

As we shall see later, it is rather difficult to find exactly $\text{Cov}(I_i, I_j)$ for most interesting sampling designs with variable probabilities P(s). The following theorem shows that it is futile to hope to find a fixed-size design in which $\text{Cov}(I_i, I_j)$ would have a simple structure without approximations.

THEOREM 3.3 Consider a sampling design with fixed size n and put

$$d = \sum_{i=1}^{N} \pi_i(1 - \pi_i) \tag{3.13}$$

Let there exist numbers $c_1, \ldots, c_N$ such that

$$\pi_i\pi_j - \pi_{ij} = c_ic_j \qquad 1 \le i \ne j \le N \tag{3.14}$$

Then, if $d \ge 3/4$, (3.14) is fulfilled with

$$c_i = \frac{1}{2}C - \sqrt{\frac{1}{4}C^2 - \pi_i(1 - \pi_i)} \qquad 1 \le i \le N \tag{3.15}$$

where C is a solution of

$$(N - 2)C = \sum_{i=1}^{N} \sqrt{C^2 - 4\pi_i(1 - \pi_i)} \tag{3.16}$$

Further, for each $d \ge 3/4$, we have

$$\left(1 - \frac{1}{4d - 1}\right)\frac{\pi_i(1 - \pi_i)\pi_j(1 - \pi_j)}{d} \le c_ic_j$$
$$\le \frac{\pi_i(1 - \pi_i)\pi_j(1 - \pi_j)}{d}\left(1 + \frac{1}{4d - 2}\right)^2 \qquad 1 \le i \ne j \le N \tag{3.17}$$

Proof. Denoting $C = \Sigma_{i=1}^{N} c_i$, (3.12) and (3.14) entail

$$\pi_i(1 - \pi_i) = c_i(C - c_i) \qquad 1 \le i \le N \tag{3.18}$$

It is easily seen that we must have $C \ne 0$ and all the numbers c_i of the same sign as C. We investigate the case $C > 0$ first: the quadratic equation (3.18) has two roots

$$c_i = \frac{1}{2}C \pm \sqrt{\frac{1}{4}C^2 - \pi_i(1 - \pi_i)} \qquad 1 \le i \le N \tag{3.19}$$

The sign +, however, is applicable in a single case at most. If $(1/4)C^2 - \pi_i(1 - \pi_i) = 0$, both signs give the same result, whereas if $(1/4)C^2 - \pi_i(1 - \pi_i) > 0$, the sign + leads to the inequality $c_i > (1/2)C$. Should this inequality hold true in more than a single case, say for units i and j, we would get the inequality $C > c_i + c_j > C$, which is a contradiction. If, however, for some i,

$$c_i = \frac{1}{2}C + \sqrt{\frac{1}{4}C^2 - \pi_i(1 - \pi_i)} \tag{3.20}$$

and for the remaining subscripts

$$c_j = \frac{1}{2}C - \sqrt{\frac{1}{4}C^2 - \pi_j(1 - \pi_j)} \qquad j \neq i$$

then, making use of

$$\sqrt{\frac{1}{4}C^2 - \pi_j(1 - \pi_j)} \leq \frac{1}{2}C - \frac{\pi_j(1 - \pi_j)}{C}$$

we get

$$c_j \geq \frac{\pi_j(1 - \pi_j)}{C} \qquad j \neq i$$

and, summing up with respect to all $j \neq i$,

$$C - c_i \geq \frac{d - \pi_i(1 - \pi_i)}{C} \geq \frac{d - 1/4}{C} \tag{3.21}$$

with d given by (3.13). Using the assumption $d \geq 3/4$, we obtain the inequality $C - c_i \geq 1/2C$. Moreover, from (3.18) it follows that $C - c_i = \pi_i(1 - \pi_i)/c_i \leq 1/4c_i$, hence $c_i \leq (1/2)C$, which contradicts (3.20). The sign + is thus inapplicable not even in a single case, provided that $d \geq 3/4$. Consequently, we obtain first that the numbers c_i are determined by the equations (3.15), which in their sum entail (3.16) to be solved for C. As a second consequence, we get that the inequality (3.21) holds true for all i, $1 \leq i \leq N$. Hence we obtain, taking (3.18) into account,

$$c_i = \frac{\pi_i(1 - \pi_i)}{C - c_i} \leq \frac{1}{4}\,\frac{C}{d - 1/4} = \frac{C}{4d - 1} \qquad 1 \leq i \leq N \tag{3.22}$$

as well as the following inequalities:

$$C - c_i \geq C - \frac{C}{4d - 1} = C\,\frac{4d - 2}{4d - 1} \qquad 1 \leq i \leq N$$

$$c_iC \geq \pi_i(1 - \pi_i) = c_i(C - c_i) \geq c_iC\,\frac{4d - 2}{4d - 1} \qquad 1 \leq i \leq N$$

$$C^2 \geq d \geq C^2 \frac{4d - 2}{4d - 1}$$

$$\pi_i(1 - \pi_i)\sqrt{\frac{4d - 2}{4d - 1}}\frac{1}{d} \leq c_i \leq \frac{\pi_i(1 - \pi_i)}{\sqrt{d}}\frac{4d - 1}{4d - 2} \qquad 1 \leq i \leq N$$

The last inequalities lead to (3.17).

It remains to investigate the case $C < 0$, which, however, can not provide any new values of $c_i c_j$. That is to say, if $\pi_i\pi_j - \pi_{ij} = d_i d_j$ for some $d_1 < 0, \ldots, d_N < 0$, then we have also $\pi_i\pi_j - \pi_{ij} = (-d_i)(-d_j)$, i.e., the numbers $-d_1 > 0, \ldots, -d_N > 0$ represent a positive solution. As shown above, the positive solution is unique; hence it must be $d_i = -c_i$ and $d_i d_j = c_i c_j$. The proof is finished.

If d is large, it follows from the inequalities (3.17) that we may write approximately

$$c_i c_j \sim \frac{\pi_i(1 - \pi_i)\pi_j(1 - \pi_j)}{d} \qquad 1 \leq i \neq j \leq N \qquad (3.23)$$

If the inclusion probabilities π_i are far from the number 1, then d will be only a bit less than the sample size n. If $\pi_i \leq \varepsilon$, $1 \leq i \leq N$, then (3.11) implies

$$n \geq d \geq n(1 - \varepsilon)$$

Roughly speaking, (3.23) is valid for large samples. For evaluating the constants c_i it is favorable to expect them in the form

$$c_i = \frac{\pi_i(1 - \lambda_i)}{[\sum_{j=1}^N \pi_j(1 - \lambda_j)]^{1/2}} \qquad 1 \leq i \leq N$$

with suitable constants λ_i. From (3.18) it follows that

$$(1 - \lambda_i)\left[1 - \frac{\pi_i(1 - \lambda_i)}{\sum_{j=1}^N \pi_j(1 - \lambda_j)}\right] = 1 - \pi_i \qquad 1 \leq i \leq N \qquad (3.24)$$

Hence, $1 - \lambda_i$ can be obtained (fairly quickly) by iterating the relation

$$1 - \lambda_i = (1 - \pi_i)\left[1 - \frac{\pi_i(1 - \lambda_i)}{\sum_{j=1}^{N} \pi_j(1 - \lambda_j)}\right]^{-1} \qquad 1 \le i \le N \quad (3.25)$$

Inserting $1 - \lambda_i = 1 - \pi_i$ on the right-hand side, the first iteration yields

$$1 - \lambda_i \doteq (1 - \pi_i)\left[1 - \frac{\pi_i(1 - \pi_i)}{d}\right]^{-1} \qquad 1 \le i \le N \quad (3.26)$$

If not satisfied with these approximate values, we can insert them again in the right-hand side of (3.25) and repeat this process till the results become stable.

When the inclusion probabilities π_i are small, we can put

$$1 - \frac{\pi_i(1 - \pi_i)}{\sum_{j=1}^{N} \pi_j(1 - \pi_j)} \doteq 1 - \frac{\pi_i}{\sum_{j=1}^{N} \pi_j} = 1 - \frac{1}{n}\pi_i \qquad 1 \le i \le N$$

Inserting this approximation in (3.26), we get

$$1 - \lambda_i \doteq 1 - \frac{n - 1}{n}\pi_i \qquad 1 \le i \le N \quad (3.27)$$

which often represents a satisfactory compromise between simplicity and precision.

Remark 3.1 Given arbitrary inclusion probabilities $\pi_1, \ldots, \pi_N$, there may exist no sampling design such that

$$\pi_i\pi_j - \pi_{ij} = c_i c_j$$

as shown by an example in Problem 3.1. It can be conjectured that for all d sufficiently large, say $d \ge 1.6$, some design satisfying (3.14) always exists; this conjecture, however, has not been proved yet. The bound 1.6 for d has been inspired by the inequality (3.17). In fact, $\pi_i\pi_j - \pi_{ij} = c_i c_j$ implies $c_i c_j \le \pi_i\pi_j$ and the latter inequality is guaranteed [according to Eq. (3.17)] for $d \ge 1.6$. In Part II, instead of trying to find a sampling design satisfying (3.14) exactly--and such a design is no doubt a complicated one--we shall investigate designs satisfying (3.14) approximately and being simple at the same time.

Remark 3.2 The relation (3.14) can be motivated by an attempt to minimize the variability of $\pi_i\pi_j - \pi_{ij} = -\text{Cov}(I_i, I_j)$ subject to (3.12). This is formulated in Problem 3.2.

Purposive sampling of a subset s_0 may be considered as degenerate probability sampling in which the probability is concentrated in one sample, $P(s = s_0) = 1$. In contradistinction to that in probability sampling we shall try to *spread* probabilities as much as is reasonable and compatible with prescribed probabilities of inclusion $\pi_1, \ldots, \pi_N$ excluding eventually those samples which are a priori undesirable. A carrier of the sampling design $\{P(s), s \subset S\}$ is defined as (any) class Ω of samples satisfying

$$P(\Omega) = \Sigma_{s\in\Omega} P(s) = 1 \tag{3.28}$$

Thus $P(s) = 0$ for all s not belonging to Ω. The smallest carrier consists, evidently, of all samples with positive probability $\Omega = \{s: \ P(s) > 0\}$.

Now we shall define a broad class of sampling designs. To each vector $p = (p_1,\ldots,p_N)$ with $0 \le p_i \le 1$ for all $1 \le i \le N$, we assign the sampling plan

$$Q_p(s) = \prod_{i\in s} p_i \prod_{i\in r} (1 - p_i) \qquad s \in S \tag{3.29}$$

where $r = S - s$ is the complement of the sample. The carrier of the design $Q_p(\cdot)$ is the class of all 2^N samples and, obviously,

$$\sum_{s\subset S} Q_p(s) = \prod_{i=1}^{N} (p_i + 1 - p_i) = 1$$

In Chapter 6, the design (3.29) will be called *Poisson sampling*. From the sampling design (3.29) further designs can be derived by means of conditioning. Let Ω be an arbitrary class of samples with $Q_p(\Omega) > 0$; put

$$Q_p(s|\Omega) = \frac{Q_p(s)}{Q_p(\Omega)} \qquad s \in \Omega \tag{3.30}$$

The sampling design (3.30) will be called *conditional Poisson sampling*; obviously, its carrier is Ω itself. Outside the carrier we put $Q_p(s|\Omega) = 0$.

To each design of the form (3.30) there correspond some inclusion probabilities

$$\pi_i = \pi_i(p,\Omega) = \sum_{s \ni i} Q_p(s|\Omega) \qquad 1 \le i \le N \tag{3.31}$$

The dependence of the vector $(\pi_1,\ldots,\pi_N)$ on the vector p and on the carrier Ω will be rather complicated in general; nevertheless, it can be deduced in practically all important cases. It is an open question whether it is possible to each vector $(\pi_1,\ldots,\pi_N)$, $0 \le \pi_i \le 1$, $1 \le i \le N$, realizable on the carrier Ω, to find a vector p such that (3.31) holds. By the realizability of $(\pi_1,\ldots,\pi_N)$, we mean that there exists a sampling design with carrier Ω such that $\pi_i = \Sigma_{s \ni i} P(s)$, $1 \le i \le N$. All the vectors realizable on a certain carrier obviously form a convex set in R^N. We conjecture that the question just formulated can be answered affirmatively. For carriers consisting of samples of fixed size, the conjecture has been proved by Dupačová (1979). Before we formulate the next theorem, it is necessary to introduce a measure of spread of the sampling design $P(\cdot)$. An appropriate measure of spread is the entropy

$$e = -\sum_{s \subset S} P(s) \ln P(s) \tag{3.32}$$

THEOREM 3.4 The entropy of any design with carrier Ω and inclusion probabilities (3.31) is less than or equal to the entropy of the design (3.29).

Proof. First assume that Ω consists of all subsets of S. Making use of the one-to-one correspondence between the sample s and the vector of inclusion indicators

$$s \leftrightarrow [I_1(s),\ldots,I_N(s)]$$

we may write $P(s) = P(I_1,\ldots,I_N)$, where $I_i = 1$ if $i \in s$, and $I_i = 0$ in the opposite case. To maximize the entropy of a sampling design is thus the same as to maximize the entropy of the joint distribution of indicators I_i, $1 \le i \le N$. Now if $\pi_1, \ldots, \pi_N$ are fixed, the marginal distributions of indicators I_i are fixed also, since they are zero-one variables and the π_i values are their expectations. Denoting the marginal distribution of I_i as $P_i(\cdot)$, we may write explicitly

$$P_i(I_i) = \pi_i^{I_i}(1 - \pi_i)^{1-I_i} \qquad I_i = 0 \text{ or } 1 \tag{3.33}$$

As is well known, the entropy of a joint distribution with fixed marginal distributions is maximal in the case of independence. If, however, the indicators are independent, then

$$P(s) = P(I_1,\ldots,I_N) = \prod_{i=1}^{N} P_i(I_i) = \prod_{i=1}^{N} \pi_i^{I_i}(1 - \pi_i)^{1-I_i}$$

$$= \prod_{i\in s} \pi_i \prod_{i\in r} (1 - \pi_i) \tag{3.34}$$

where s is the sample consisting of those units i for which $I_i = 1$. Obviously, the sampling design (3.34) is identical with Poisson sampling (3.29), where we put $p_i = \pi_i$.

Thus we have proved--for Ω consisting of all subsets of S--not only the assertion of the theorem, but moreover that the inclusion probabilities connected with the design (3.29) are equal to π_i, i.e., that

$$\pi_i = \sum_{s\ni i} Q_p(s) = p_i \qquad 1 \le i \le N \tag{3.35}$$

Now consider an arbitrary carrier Ω. If there existed a design $P(\cdot)$ having the same probabilities of inclusion and the same carrier as $Q_p(\cdot|\Omega)$, and a larger entropy, it would mean that the design

$$P^*(s) = \begin{cases} P(s)Q_p(\Omega) & \text{if } s \in \Omega \\ Q_p(s) & \text{if } s \notin \Omega \end{cases} \tag{3.36}$$

would have the same probabilities of inclusion as the design $Q_p(s)$, and at the same time a larger entropy. In fact, we have

$$\pi_i = \sum_{s \ni i} P^*(s) = Q_p(\Omega) \sum_{s \ni i} P(s) + \sum_{\substack{s \ni i \\ s \notin \Omega}} Q_p(s)$$

$$= Q_p(\Omega) \sum_{s \ni i} Q_p(s/\Omega) + \sum_{\substack{s \ni i \\ s \notin \Omega}} Q_p(s)$$

$$= \sum_{\substack{s \ni i \\ s \in \Omega}} Q_p(s) + \sum_{\substack{s \ni i \\ s \notin \Omega}} Q_p(s) = \sum_{s \ni i} Q_p(s) = p_i \qquad 1 \le i \le N$$

and further,

$$-\sum P^*(s) \ln P^*(s) = -\sum_{s \in \Omega} P(s)Q_p(\Omega) \ln [P(s)Q_p(\Omega)]$$

$$-\sum_{s \notin \Omega} Q_p(s) \ln Q_p(s) = -Q_p(\Omega) \sum_{s \in \Omega} P(s) \ln P(s)$$

$$- Q_p(\Omega) \ln Q_p(\Omega) - \sum_{s \notin \Omega} Q_p(s) \ln Q_p(s)$$

$$> -Q_p(\Omega) \sum_{s \in \Omega} Q_p(s/\Omega) \ln Q_p(s/\Omega) - Q_p(\Omega) \ln Q_p(\Omega)$$

$$-\sum_{s \notin \Omega} Q_p(s) \ln Q_p(s) = -\sum_{s \subset S} Q_p(s) \ln Q_p(s)$$

This is, however, a contradiction with the first part of the proof; the theorem is proved in its full extent.

Remark 3.3 We shall see later that the relation (3.35) retains its *approximate* validity for some important carriers. For those carriers we have

$$\pi_i(p,\Omega) \sim p_i \qquad 1 \le i \le N \tag{3.37}$$

PROBLEMS AND NOTES

3.1 Consider a population of size N = 4 and a sampling design of size n = 2 with probabilities $\pi_1 = \pi_2 = 1 - \varepsilon$, $\pi_3 = \pi_4 = \varepsilon$, where $0 < \varepsilon \le 1/2$. Prove that (3.14) implies

$$\pi_i \pi_j - \pi_{ij} = \frac{1}{3}\varepsilon(1 - \varepsilon) \qquad 1 \le i \ne j \le 4$$

which yields a contradiction (since it entails $\pi_{34} < 0$) if $\varepsilon < 1/4$. Thus, for the vector of inclusion probabilities $(1 - \varepsilon, 1 - \varepsilon, \varepsilon, \varepsilon)$

with $\varepsilon < 1/4$, there exists no design satisfying (3.14).

3.2 Given a fixed sample size n and given inclusion probabilities $\pi_1, \ldots, \pi_N$, $\Sigma_{i=1}^{N} \pi_i = n$, then the sampling design satisfying (3.14) maximizes

$$-\sum_{i \neq j}^{N} \sum^{N} (\pi_i \pi_j - \pi_{ij}) \ln (\pi_i \pi_j - \pi_{ij}) \tag{3.38}$$

within the family of all designs of fixed sample size n with fixed probabilities π_i and such that $\pi_i \pi_j - \pi_{ij} \geq 0$, $1 \leq i \neq j \leq N$. Thus, such a design makes the covariances $\text{Cov}(I_i, I_j)$ most uniform.

We shall see that a design maximizing (3.32) need not necessarily maximize (3.38), although it will do so approximately.

3.3 Consider a design of fixed sample size n = 2, given by formula

$$P(s) = \frac{2\alpha_i \alpha_j}{1 - \sum_{k=1}^{N} \alpha_k^2} \qquad \text{for } s = \{i,j\} \qquad 1 \leq i \neq j \leq N \tag{3.39}$$

where α_i are positive numbers, such that $\Sigma\, \alpha_i = 1$. Prove that arbitrary inclusion probabilities $\pi_i > 0$, $\Sigma_{i=1}^{N} \pi_i = 2$, can be realized choosing the α_i values properly, and that for these α_i we have

$$\pi_i = \frac{2\alpha_i (1 - \alpha_i)}{1 - \sum_{k=1}^{N} \alpha_k^2} \qquad 1 \leq i \leq N \tag{3.40}$$

Further, prove that $P(\cdot)$ maximizes entropy in the family of designs of sample size 2 and with inclusion probabilities (3.40).

3.4 Prove the following relations for inclusion probabilities π_i, π_{ij} connected with conditional Poisson sampling $Q_p(s|\Omega)$ given by (3.30);

$$\begin{aligned} \pi_i &= p_i \frac{Q_p(\Omega | I_i = 1)}{Q_p(\Omega)} \\ \pi_{ij} &= p_i p_j \frac{Q_p(\Omega | I_i = I_j = 1)}{Q_p(\Omega)} \end{aligned} \tag{3.41}$$

4

MORE ABOUT INFERENCE

In connection with probability sampling, we can regard $y_1, \ldots, y_N$ as unknown *parameters*, or, equivalently, as random variables whose distribution law is totally unknown. Having selected a sample $s = \{i_1,\ldots,i_n\}$ and established the corresponding y values, i.e., $y_{i_1}, \ldots, y_{i_n}$, let us denote the total observation by (s,y) or more briefly by ω:

$$\omega = (s,y) = (i_1,\ldots,i_n; y_{i_1},\ldots,y_{i_n}) \tag{4.1}$$

The indices $i_1, \ldots, i_n$ will symbolize all useful facts known about the units--for example, the values of auxiliary variables whose population totals are known.

Perhaps the most important aim of statistical inference in sample surveys is finding a pivotal quantity

$$t = \frac{Y - \hat{Y}(\omega)}{v(\omega)} \tag{4.2}$$

with the following properties:

1. $\hat{Y}(\omega)$ and $v(\omega)$ depend only on the observation.
2. The distribution of t is as independent of $(y_1,\ldots,y_N)$ as possible and is approximately normal N(0,1).
3. $Ev(\omega)$ is as small as possible.

It is known that for satisfaction of these properties it is beneficial to spread the probabilities of the sampling design $\{P(s), s \subset S\}$ as much as possible, i.e., roughly, to maximize the entropy. It would be nice to have a more formal expression of this fact, but we presently have not.

If the distribution of $(y_1,\ldots,y_N)$ is not completely unknown, as in the probability speculation approach described in Chapter 2, then property 2 may be replaced as follows:

2'. Superimposing the prior distribution of $(y_1,\ldots,y_N)$ with the employed probability sampling, the resulting distribution of t in the combined experiment is independent as possible of the choice of the prior from the family of admissible priors and is approximately normal N(0,1).

The extreme kind of speculation is the Bayesian approach in which we assume the distribution of $(y_1,\ldots,y_N)$ completely known.

Although random sampling is a great achievement in the history of statistics, it does not represent a pure bonus standing above all criticism. The most important objection may be formulated as follows: What relevance have samples that could have been drawn and their probabilities, if we know that they have not been drawn? Should not inferences be based just on the particular sample that has been drawn?

The above criticism is serious but not damaging. It forces the inference in the purely Bayesian realm; however, Bayesian methods are helpless if the prior probability description of the population is too vague. The purpose of randomization is to achieve simplicity in the behavior of the pivotal quantity t even in complicated situations. It remains to be added that any component of randomization that does not serve this aim is senseless and should not appear in our inferences. To illustrate the last point imagine that the sampling experiment is carried out in two stages: in the first stage we select one of M sampling rules $P_j(s)$, $1 \le j \le M$, with probabilities $\alpha_1, \ldots, \alpha_M$; in the second stage we draw a sample according to the selected sampling rule. Obviously, the probabilities corresponding to the resulting combined sampling rule equal

$$P(s) = \sum_{j=1}^{M} \alpha_j P_j(s) \tag{4.3}$$

and a serious question arises: Should we base our inference on probabilities $P(s)$ or on $P_j(s)$, where j is the label selected in the first stage of the sampling experiment? The answer is that we should prefer $P_j(s)$, provided that there is a pivotal quantity which does not behave any worse under P_j than the pivotal quantity under P; otherwise we should prefer $P(s)$.

Example 4.1 Simple random sampling of size n is defined by

$$P(s) = \begin{cases} \binom{N}{n}^{-1} & \text{if size } s = n \\ 0 & \text{otherwise} \end{cases}$$

If $P_1, \ldots, P_M$ in (4.3) were simple random samplings of varying sizes $n_1, \ldots, n_M$, their mixing would exemplify an inflation or randomness, because the pivotal quantity (4.2), where $\hat{Y}$ and $v(\omega)$ are defined by (2.6) and (2.9), applied to particular simple random samplings does not behave any worse than any pivotal quantity for combined sampling. As a matter of fact, the best pivotal quantity for the combined sampling is obtained just if the pivotal quantities for P_j are used in case the sample size is n_j.

PARAMETRIC AND SAMPLE SPACES In probability sampling $(y_1,\ldots,y_N)$ is the unknown parameter, and therefore, the N-dimensional Euclidean space R^N may serve as the *parametric space*.

The points ω of the sample space have been introduced in (4.1). Denoting $y_{i_k} = v_k$ and $\omega = (i_1,\ldots,i_n,v_1,\ldots,v_n)$, the sample space will be a union of 2^N spaces (s,R^{n_s}), each labeled by a sample s, $s \subset S$, and of dimension n_s = size s:

$$\text{Sample space} = \bigcup_{s\subset S} (s,R^{n_s}) \tag{4.4}$$

The combined parametric-and-sample space thus is a union of of spaces (s,R^{N+n_s}), $s \subset S$. For each s let the corresponding σ fields be the usual Borel fields.

DISTRIBUTION ON THE PARAMETRIC-AND-SAMPLE SPACE This distribution will be determined by a prior distribution of $y = (y_1,\ldots,y_N)$ in R^N and by probability sampling $P(s)$. Probability sampling entails that for any point $\omega = (i_1,\ldots,i_{n_s},v_1,\ldots,v_{n_s})$ the following conditional probability holds:

$$P(\omega|y) = \begin{cases} P(s) & \text{if } v_k = y_{i_k} \quad 1 \le k \le n_s \\ 0 & \text{otherwise} \end{cases} \tag{4.5}$$

This corresponds to the case in which the y_i terms are observed without errors. If there are nonsampling errors, then $P(\omega \in A|y)$ is a distribution on the sample space (4.4) such that

$$P(\omega \in (s,R^{n_s})|y) = P(s|y) = P(s) \tag{4.6}$$

independently of y.

LIKELIHOOD FUNCTION As we can see from (4.5), the likelihood equals

$$\ell(y|\omega) = \begin{cases} \lambda(\omega)P(s) & \text{if } y_{i_k} = v_k \quad 1 \le k \le n_s \\ 0 & \text{otherwise} \end{cases} \tag{4.7}$$

where $\lambda(\omega)$ may be any function of ω. Choosing $\lambda(\omega) = 1/P(s)$, we obtain

$$\ell(y|\omega) = \begin{cases} 1 & \text{if } y_{i_k} = v_k \quad 1 \le k \le n_s \\ 0 & \text{otherwise} \end{cases} \tag{4.8}$$

Consequently the likelihood function does not depend on the applied probability sampling. Therefore, if adhering to the likelihood principle, we should ignore the way the sample was selected. The likelihood principle is satisfied, for example, if we apply Bayesian inference.

BAYESIAN INFERENCE The basic feature of Bayesian inference in probability sampling is the fact that the posterior distribution of $(y_1,\ldots,y_N)$ given the observation ω does not involve the probabilities $P(s)$, $s \subset S$. It does not depend either on how the sample obtained was selected or how probable were the alternative samples. The conditional distribution of $(y_1,\ldots,y_N)$ may be fully described by giving the conditional distribution of the variables y_i for $i \in r = S - s$ given those in the sample. For example, if $s = \{1,2,\ldots,n\}$ and the a priori distribution of $(y_1,\ldots,y_N)$ is described by a density $f(y_1,\ldots,y_N)$, then the posterior (conditional) distribution of $(y_{n+1},\ldots,y_N)$ given $y_1 = v_1, \ldots, y_n = v_n$ equals

$$g(y_{n+1},\ldots,y_N|v_1,\ldots,v_n) = \frac{f(v_1,\ldots,v_n,y_{n+1},\ldots,y_N)}{\int\cdots\int f(v_1,\ldots,v_n,y_{n+1},\ldots,y_N)\, dy_{n+1}\cdots dy_N} \tag{4.9}$$

as is well known. Thus the Bayesian inference in probability sampling does not differ much from Bayesian inference in general.

Example 4.2 Assume that conditionally on (μ,h) the y_i terms are independent normal random variables with expectation μ and variance $1/h$. Furthermore, let (μ,h) also be random with improper prior density

$$\phi(\mu,h) = h^{-1} \qquad -\infty < \mu < \infty \qquad h > 0 \tag{4.10}$$

We call ϕ improper since its integral over the region $(-\infty < \mu < \infty, h > 0)$ is infinite. However, in Bayesian statistics this density is often used to express a vague prior opinion concerning (μ,h). Of course, the joint density of $(y_1,\ldots,y_N,\mu,h)$ will also be improper, but conditional densities will be integrable after all. A realistic interpretation of ϕ is as of a limit of proper prior densities, even if this point is not completely clear as we shall see in Remark 4.1, below.

The joint density of $(y_1,\ldots,y_N,\mu,h)$ is as follows:

$$g(y_1,\ldots,y_N,\mu,h) = ch^{(1/2)N-1} \exp\left[-\frac{1}{2}\sum_{i=1}^{N}(y_i - \mu)^2 h\right] \tag{4.11}$$

Introducing

$$Y_r = \sum_{i=n+1}^{N} y_i \tag{4.12}$$

the joint density of $(y_1,\ldots,y_n,Y_r,\mu,h)$ will be

$$g(y_1,\ldots,y_n,Y_r,\mu,h)$$
$$= ch^{(1/2)n-(1/2)} \exp\left\{-\frac{1}{2}h \sum_{i=1}^{n} (y_i - \mu)^2 - \frac{1}{2}\frac{h}{N-n}[Y_r - (N-n)\mu]^2\right\} \tag{4.13}$$

Integrating out (μ,h) we obtain

$$f(y_1,\ldots,y_n,Y_r) = \int_0^{\infty}\int_{-\infty}^{\infty} g(y_1,\ldots,y_n,Y_r,\mu,h)\, d\mu\, dh$$
$$= c\left[\sum_{i=1}^{n} (y_i - \bar{y})^2 + \frac{(\hat{Y} - Y)^2 n}{N(N-n)}\right]^{-(1/2)n} \tag{4.14}$$

where $\bar{y} = (1/n)\,\Sigma_{i=1}^{n}\, y_i$ and $\hat{Y} = N\bar{y}$. Consequently, we can see that the conditional distribution given $y_1, \ldots, y_n$ of the ratio

$$t = \frac{\hat{Y} - Y}{\left[\frac{N(N-n)}{n(n-1)} \sum_{i=1}^{n} (y_i - \bar{y})^2\right]^{1/2}} \tag{4.15}$$

is the Student distribution with n - 1 degrees of freedom.

Remark 4.1 A proper prior density for (μ,h) is considered in Problem 4.2. If we regard (4.10) as a limit case of the prior of Problem 4.2 and pass to the limit in the expression for t, we do not obtain exactly (4.15) but (4.15) with n - 1 replaced by n. In the situation considered in Problem 4.2, $\hat{Y}$ is replaced by a weighted average of $\hat{Y}$ and a prior value $\underline{a}$, and the variance is estimated as a weighted average of $\Sigma_{i=1}^{n}\,(y_i - \bar{y})^2$, $(\bar{y} - a)^2$, and of the prior value β.

Remark 4.2 A similar result could be formulated for an arbitrary sample. Then $\Sigma_{i=1}^{n}\,(y_i - \bar{y})^2$ would be replaced by $\Sigma_{i\in s}\,(y_i - \bar{y})^2$ in (4.15).

ADMISSIBILITY Besides the Bayesian approach there is another group of results that seem to undermine the postulate of randomness in the selection of a sample. These results are concerned with admissibility and are due to Godambe and Yoshi (see Yoshi, 1969, for a review). We say that an estimate $e(\omega)$ of Y is admissible under a sampling design $\{P(s),\ s \subset S\}$ if there does not exist any other estimate $e^*(\omega)$ such that for all points $(y_1,\ldots,y_N)$ the inequality

$$\sum_S P(s)[e^*(\omega) - Y]^2 \le \sum_S P(s)[e(\omega) - Y]^2$$

holds with a sharp inequality for at least one point. It was proved, for example, that the simplest estimator $Y = N\bar{y}$, where $\bar{y}$ denotes the sample average, is admissible for Y under *any* sample design. This seems to indicate that our inference may be largely independent of the sample design. However, a careful analysis of the proofs of the admissibility results shows that jumping to such a conclusion would be futile.

For example, in the proof of the admissibility of $N\bar{y}$ the basic idea is the assumption that the y_i terms are independent random variables with a common distribution $N(0,\sigma^2)$ where $\sigma^2 \to \infty$. Now if a hypothesis like this is a priori acceptable, we should use the probability speculation approach giving us $N\bar{y}$ as the best estimate for any sample and selection procedure (see Example 2.1). But if this hypothesis is doubtful, the usefulness of admissibility of $N\bar{y}$ will be doubtful also, since it does not apply if we exclude populations compatible with the above model. Thus admissibility results have a useful content only for well-mixed populations.

LINEARITY OF ESTIMATES Linearity of estimates for Y in sample surveys is defined in a slightly unusual way. A general linear estimate $\hat{Y}$ has the form

$$\hat{Y} = \sum_{i \in s} y_i w_i(s) \qquad (4.16)$$

where the coefficients $w_i(s)$ depend on the sample s. The sample s here represents all available auxiliary information we may have about the units in s. For example, in ratio estimation

$$\hat{Y} = \frac{X}{x} y \tag{4.17}$$

where x and y are sample totals. Thus, for (4.17),

$$w_i(s) = \frac{X}{x}$$

Linearity in the classical sense would be connected with a sequence of observations $y_{i_1}, \ldots, y_{i_n}$ drawn in a succession and combined as follows:

$$Y = \sum_{k=1}^{n} c_k y_{i_k} \tag{4.18}$$

In contradistinction to that, in our framework, there is only an *unordered subset* of units, the constants $w_i(s)$ are attached to units and not to the order of appearance, and they depend on s.

REPRESENTATIVENESS Representativeness is a joint property of an estimator $e = e(\omega)$ and a sampling design $P = \{P(s), s \subset S\}$. It was introduced in Hájek (1959) but its role has not been duly appreciated in the literature. We say that the pair (e,P) is representative with respect to numbers x_i if

$$P(e(s,x) = X) = 1 \tag{4.19}$$

i.e., if the estimate applied to the population $(x_1,\ldots,x_N)$ estimates the total without error with probability 1. Representativeness is important as one of conditions of optimality and its presence also enables us to simplify the expression for the mean square error (see Chapter 16).

GUESSING THE PARAMETERS The precision may be often improved if some parameters of the model are guessed a priori instead of being estimated from the sample. For example, if we expect linear regression of the y_i terms on the x_i terms, i.e.,

$$My_i = \alpha + \beta x_i \tag{4.20}$$

then it is often possible to guess that the regression passes through the origin, i.e., that $\alpha = 0$ or nearly so. Then, estimates based on the simplified model $My_i = \beta x_i$ will be more precise than those based on the original model. Moreover, we can use the fact that

$$\beta = \frac{Y}{X} \tag{4.21}$$

and estimate it by y/x rather than by the more complicated estimator. Of course, errors in the assumption $\alpha = 0$ may bring a negative effect if α differs from 0 widely.

ERROR ESTIMATION It is usual to estimate the square error $(e - Y)^2$ by the *mean square error*

$$E(e - Y)^2 = \Sigma_s P(s)[e(\omega) - Y]^2 \tag{4.22}$$

and then to estimate the mean square error $E(e - Y)^2$ by some estimate $v(\omega)$ such that

$$Ev(\omega) = E(e - Y)^2 \tag{4.23}$$

Thus the connection between the actual square error $(e - Y)^2$ and $v(\omega)$ is mediated by $E(e - Y)^2$. This standard procedure may lead to undesirable situations, in which $v(\omega)$ tends to be large, when we may expect $(e - Y)^2$ to tend to be below the average. Keeping this in mind we shall try to keep $v(\omega)$ close not to $E(e - Y)^2$ but to $(e - Y)^2$ itself, i.e., to minimize

$$E[v(\omega) - (e - Y)^2]^2 \tag{4.24}$$

Example 4.3 If mixing M simple random samplings with weights $\alpha_1, \ldots, \alpha_M$ as in Example 4.1, the estimate $N\bar{y}$, where $\bar{y}$ is the sample average, will be unbiased and have the variance

$$\mathrm{Var}(N\bar{y}) = \sum_{j=1}^{M} \alpha_j \frac{N^2}{n_j} \frac{N - n_j}{N - 1} \sigma^2 \tag{4.25}$$

where n_j are sizes of respective simple random samplings and σ^2 is given by (1.6). A good estimate of $\mathrm{Var}(N\bar{y})$ would be

$$v_0(\omega) = \sum_{j=1}^{M} \alpha_j \frac{N}{n_j} \frac{N - n_j}{n - 1} \sum_{i \in s} (y_i - \bar{y})^2 \tag{4.26}$$

where n = size s.

If we try to minimize (4.24), however, a better estimate is

$$v_1(\omega) = \frac{N}{n} \frac{N - n}{n - 1} \sum_{i \in s} (y_i - \bar{y})^2 \tag{4.27}$$

A MORE COMPREHENSIVE DEFINITION OF A SAMPLE When convenient a sample may be defined in a more comprehensive way. Imagining, for example, that the units are drawn one by one until the sampling is finished according to some stopping rule, we may define a sample as a finite *sequence* of units. This sequence distinguishes the *order* of units, may be of *variable length*, and may include one unit several times (*multiplicities*). If we denote this new concept of a sample as s', we may write

$$s' = (i_1, \ldots, i_m)$$
$$\text{where } m = 0, 1, 2, \ldots \qquad 1 \le i_k \le N \qquad k = 1, 2, \ldots, m \tag{4.28}$$

indicating by $(\cdot)$ that it is a sequence and not a subset.

Obviously, to every sample s' there corresponds a sample s defined as a subset of those units which appear in s' at least once:

$$s' \to s \tag{4.29}$$

Example 4.4 If s' = (16,4,8,4,16,16,7), then s = {4,7,8,16}, m = 7, n = 4.

Denote the class of all finite sequences s' over S by ζ'. Then a more comprehensive sampling design will be given by a probability distribution $P'(s')$, $s' \in \zeta'$. This distribution generates a distribution $P(s)$, $s \subset S$, where s corresponds to s':

$$P(s) = \sum_{s' \to s} P'(s') \tag{4.30}$$

i.e., we sum up over all s' that are projected to s; there are infinitely many such s'.

Example 4.5 For s = {1,2} the following samples s' project into s:

(1,2)	(1,2,2)	(2,1,1)	(1,2,2,2)	etc.
(2,1)	(2,1,2)	(1,2,1)	(2,1,2,2)	
	(2,2,1)	(1,1,2)	(2,2,1,2)	
			(2,2,2,1)	

SUFFICIENCY If we know s and y_i for $i \in s$, then we can reconstruct the observation for any s' such that $s' \to s$, because in s' appear only those units that are already in s. Thus the knowledge of $\omega = (s,y)$ is sufficient for statistical inference, and so we are fully justified in introducing the sample as a subset. To any estimator based on $(s',y) = (i_1,\ldots,i_m,y_{i_1},\ldots,y_{i_m})$ there is a corresponding estimator defined as its conditional expectation given $s' \to s$:

$$e(s,y) = \frac{\Sigma_{s'\to s}\, e'(s',y)P'(s')}{\Sigma_{s'\to s}\, P'(s')} \tag{4.31}$$

It can be shown that

$$Ee(s,y) = Ee'(s,y) \tag{4.32}$$

and

$$\text{Var } e(s,y) \le \text{Var } e'(s',y) \tag{4.33}$$

where the inequality is sharp unless e' = e. The sufficiency of (s,y) can be proved to hold also in the usual technical sense, namely that the conditional probability of (s',y) given the corresponding (s,y) is independent of y:

$$P((s',y)\,|\,(s,y)) = \frac{P'(s')}{\Sigma_{s''\to s}\, P'(s'')} \qquad \text{if } s' \to s \tag{4.34}$$

PROBLEMS AND NOTES

4.1 A statistic $t = t(s,y)$ from (4.15) has the same distribution under either of the following two conditions:

(a) The y_i terms are some fixed numbers, $y_i = v_i$, $1 \le i \le N$, and the sampling is simple random of size n.

(b) The sampling is arbitrary of fixed size n, but the y_i terms are independent random variables with an arbitrary common distribution F, and we consider them under the condition that for some permutation $(r_1,\ldots,r_N)$ of $(1,2,\ldots,N)$ the relation $y_{r_i} = v_i$, $1 \le i \le N$, holds.

4.2 *A continuous Bayesian model* Consider the same superpopulation model as in Example 4.2 with the modification that the prior density of (μ,h) would be

$$\phi(\mu,h) = ch^{1/2} \exp\left[-\frac{1}{2}h\alpha(\mu - a)^2\right]h^{(1/2)\nu-1} \exp\left(-\frac{1}{2}h\nu\beta\right) \tag{4.35}$$

where α, a, ν, and β are known constants. It means that μ is a priori normal $N(a, 1/h\alpha)$ and h is a priori gamma-distributed with expectation $1/\beta$ and ν degrees of freedom, and that μ is a priori independent of h. Show then that the statistic

$$t = \frac{Y - \dfrac{n(N + \alpha)}{N(n + \alpha)}\hat{Y} - \dfrac{(N - n)\alpha}{n + \alpha}a}{\left\{\left[\dfrac{(N - n)(N + \alpha)}{(n + \alpha)(\nu + n)}\right]\left[\sum_{i=1}^{n}(y_i - \bar{y})^2 + \dfrac{\alpha n}{n + \alpha}(\bar{y} - a)^2 + \nu\beta\right]\right\}^{1/2}} \tag{4.36}$$

will have the Student distribution with $\nu + n$ degrees of freedom.

If $\alpha \to 0$ and $\nu \to 0$, the t will not converge to t given by (4.15). In order to obtain (4.15) we need $\nu + n \to n - 1$ (Ericson, 1969).

4.3 *A discrete Bayesian model* Assume conditionally on $\theta = (p_1,\ldots, p_k)$ that the y_i terms are independent random variables such that $P(y_i = b_j) = p_j$, $1 \le i \le N$, $1 \le j \le k$, where $B = \{b_1,\ldots,b_k\}$ is a finite set of numbers. Furthermore, let $(p_1,\ldots,p_k)$ be a random vector having the Dirichlet distribution given by the density

$$\phi(p_1,\ldots,p_k) = \frac{\Gamma(\varepsilon)}{\Gamma(\varepsilon_1)\cdots\Gamma(\varepsilon_k)} \prod_{j=1}^{k} p_j^{\varepsilon_j - 1}$$

$$\text{if } p_1 + \cdots + p_k = 1 \qquad 0 \le p_1, \ldots, p_k \le 1 \tag{4.37}$$

where $\varepsilon_1, \ldots, \varepsilon_k$ are some a priori known positive numbers and $\varepsilon = \varepsilon_1 + \cdots + \varepsilon_k$. The density ϕ is understood with respect to the Lebesgue measure on the simplex $0 \le p_1,\ldots,p_k \le 1$, $p_1 + \cdots + p_k = 1$. Then for any vector $(y_1^0,\ldots,y_N^0)$ such that $y_i^0 \in B$, $1 \le i \le N$, we have

$$P(y_1 = y_1^0,\ldots,y_N = y_N^0 \mid \varepsilon_1,\ldots,\varepsilon_k) = \frac{\Gamma(\varepsilon)}{\Gamma(N + \varepsilon)} \prod_{j=1}^{k} \frac{\Gamma(N_j + \varepsilon_j)}{\Gamma(\varepsilon_j)} \tag{4.38}$$

where N_j denotes the number of the y_i^0 terms equal to b_j, $1 \le j \le k$. Moreover, the random variables $N_1, \ldots, N_k$ have the following expectations, variances, and covariances under (4.38):

$$M(N_j) = \frac{N\varepsilon_j}{\varepsilon} \qquad D(N_j) = \frac{N(N + \varepsilon)\varepsilon_j(\varepsilon - \varepsilon_j)}{\varepsilon^2(\varepsilon + 1)} \qquad 1 \le j \le k$$

$$\mathrm{Cov}(N_j, N_h) = -\frac{N(N + \varepsilon)\varepsilon_j\varepsilon_h}{\varepsilon^2(\varepsilon + 1)} \qquad 1 \le j \ne h \le k \tag{4.39}$$

The population total obviously equals $Y = \Sigma_{j=1}^{k} b_j N_j$ and its expectation conditional on $\omega = (s,y)$ equals, under (4.38),

$$M(Y|\omega) = \sum_{j=1}^{k} b_j \frac{(N + \varepsilon)n_j + (N - n)\varepsilon_j}{n + \varepsilon}$$

where n_j is the frequency of the value b_j in the sample s, $1 \le j \le k$. The conditional variance of Y equals

$$D(Y|\omega) = \frac{(N - n)(N + \varepsilon)}{(n + \varepsilon)(n + \varepsilon + 1)} \left\{ \sum_{j=1}^{k} b_j^2 n_j - \frac{\left(\sum_{j=1}^{k} b_j n_j\right)^2}{n} + \frac{\varepsilon n}{n + \varepsilon}\left[\sum_{j=1}^{k} b_j \left(\frac{n_j}{n} - \frac{\varepsilon_j}{\varepsilon}\right)\right]^2 + \sum_{1}^{k} b_j^2 \varepsilon_j - \frac{\left(\sum_{j=1}^{k} b_j \varepsilon_j\right)^2}{\varepsilon} \right\} \tag{4.40}$$

where $n = n_1 + \cdots + n_k$.

The above model is applied as follows: we choose the ε_j terms proportionate to a priori expected frequencies of the values b_j, $1 \le j \le k$, and the sum $\varepsilon = \varepsilon_1 + \cdots + \varepsilon_k$ is chosen the smaller the vaguer is our prior information. In the case of extreme vagueness we take the limit $\varepsilon \to 0$ (Ericson, 1969).

4.4 *Admissibility of* $N\bar{y}$ Prove a weaker and easier result than the one by Joshi and Godambe: $N\bar{y}$ is admissible in the class of *linear* estimates.

4.5 In Example 4.3 prove that

$$E[v_1(\omega) - (N\bar{y} - Y)^2]^2 < E[v_0(\omega) - (N\bar{y} - Y)^2]^2 \tag{4.41}$$

4.6 Define

$$P'(s') = \begin{cases} \left(\frac{1}{N}\right)^n & \text{if } s' \text{ contains } n \text{ distinct units and } y_{i_m} \neq y_{i_k},\ k < m \\ 0 & \text{otherwise} \end{cases} \tag{4.42}$$

Show that Eq. (4.30) yields

$$P(s) = \begin{cases} \binom{N}{n}^{-1} & \text{if size } s = n \\ 0 & \text{otherwise} \end{cases} \tag{4.43}$$

4.7 Put

$$e'(s',y) = \frac{1}{m} \sum_{k=1}^{m} y_{i_k} \tag{4.44}$$

and show that Eq. (4.31) with $P'(s')$ given by (4.42) yields, for $s' \to s$,

$$e(s,y) = \frac{1}{n} \sum_{i \in s} y_i \tag{4.45}$$

where n is the size of s.

Part II

METHODS OF SAMPLING

5

SIMPLE RANDOM SAMPLING

Simple random sampling of size n is defined by

$$P(s) = \begin{cases} \binom{N}{n}^{-1} & \text{if size } s = n \\ 0 & \text{otherwise} \end{cases} \tag{5.1}$$

It spreads the probability over the class of all samples of size n as uniformly as possible and has the maximum entropy in the family of probability samplings such that P(size s = n) = 1.

Simple random sampling may be physically realized by means of tables of random digits. We label the units by numbers 1, 2, ..., N and note how many digits the number N has. If it is, say, a four-digit number, we select from the tables a sequence of four-digit numbers representing the sample. The first number is selected at random, and we leave out all numbers greater than N and all repetitions. We continue until the sample consists of n distinct numbers, all smaller than or equal to N. Number 3 will appear as 0003, and 0000 may represent the 10000.

Example 5.1 Let N = 76, n = 16. A sequence of two-digit numbers from tables of random digits provided:

$\underline{60}$ $\underline{17}$ 97 $\underline{01}$ $\underline{09}$ 80 $\underline{51}$ $\underline{29}$ $\underline{14}$ 81
$\underline{07}$ $\underline{69}$ 79 $\underline{20}$ $\underline{35}$ 90 $\underline{45}$ $\underline{31}$ 69 $\underline{11}$
93 $\underline{19}$ $\underline{18}$

The numbers 97, 80, 81, etc. have been omitted because they are larger than 76; the number 69 was a repetition. The resulting sample consists of the underlined units:

$$s = \{1,7,9,11,14,17,18,19,20,29,31,35,45,51,60,69\}$$

If the first digit in N is 1, we may identify 1 with odd numbers 1, 3, 5, 7, 9 and 0 with even numbers 0, 2, 4, 6, 8, reducing so the number of cases when the selected number is greater than N.

Example 5.2 Let N = 120, n = 10. Tables of random digits provided the following sequence:

<u>501</u>	<u>081</u>	176	<u>803</u>	127	558	603	570
927	592	328	341	594	<u>095</u>	<u>106</u>	186
<u>436</u>	797	321	566	<u>250</u>	559	378	742
<u>867</u>	<u>607</u>	<u>248</u>					

Transforming the odd first digits into 1 and the even first digits into 0, we obtain this sample:

$$s = \{3,7,36,48,50,67,81,95,101,106\}$$

When the total frame consists of M partial frames each numbered separately, every unit can be associated with a pair of ordinal numbers--one referring to the frame and the other referring to the position within the frame. Then it is necessary to express the sample in terms of these pairs of numbers. This may be achieved by computing cumulative sizes of partial frames. If N_j is the size of the jth frame, the cumulative size of the same frame will be $N_1 + \cdots + N_j$. Having selected a unit i, we identify j such that

$$N_1 + \cdots + N_{j-1} < i \leq N_1 + \cdots + N_j \tag{5.2}$$

and then transform i into the pair

$$i \to (j,\ i - N_1 - \cdots - N_{j-1}) = (j,k) \tag{5.3}$$

which means that the unit is the kth unit in the jth frame.

Example 5.3 For each county let us have a separate list of elementary schools. The sizes as well as cumulative sizes of the counties are given in Table 5.1.

TABLE 5.1

County	Number of schools	Cumulative number of schools	Selected units	Selected units expressed as a pair
1	41	41		
2	26	67		
3	28	95	68	(3, 1)
4	15	110		
5	46	156	125	(5, 15)
6	51	207		

Assume we have selected two units bearing the numbers 68 and 125. The first of them is identical with the 1st unit in the 3rd county, the other is identical with the 15th unit in the 5th county. The ordering of pairs here is lexicographical.

We should also mention some procedures that cannot be viewed as physical realizations of sample random sampling--for example, sampling in which we first select a simple random sample of buildings and then pick up one family in every building. Here the probability of inclusion of a family is inversely proportionate to the size of the building, to say the least. The uniformity of probabilities of inclusion is also violated if we select areas of unequal size by means of selecting points in a map, or if we select persons by means of cards in a file where the number of cards per person varies.

Now let us turn to probabilities of inclusion defined in Chapter 3.

THEOREM 5.1 In simple random sampling of size n we have the following probabilities of inclusion:

$$\pi_i = \frac{n}{N} \qquad 1 \le i \le N \tag{5.4}$$

$$\pi_{ij} = \frac{n(n-1)}{N(N-1)} \qquad 1 \le i \ne j \le N \tag{5.5}$$

Proof. There are $\binom{N-1}{n-1}$ samples of size n including the unit i, since the remaining n - 1 units may be selected out of N - 1 units just in $\binom{N-1}{n-1}$ ways. Thus

$$\pi_i = \sum_{s \ni i} P(s) = \binom{N-1}{n-1}\binom{N}{n}^{-1} = \frac{n}{N}$$

Similarly there are $\binom{N-2}{n-2}$ samples of size n including both units i and j. Consequently,

$$\pi_{ij} = \sum_{s \ni i,j} P(s) = \binom{N-2}{n-2}\binom{N}{n}^{-1} = \frac{n(n-1)}{N(N-1)}$$

From the theorem it follows that the indicators of inclusion I_i, I_j are negatively correlated and

$$\text{Cov}(I_i, I_j) = \pi_{ij} - \pi_i \pi_j = \frac{n(n-1)}{N(N-1)} - \frac{n^2}{N^2} = -\frac{n(N-n)}{N^2(N-1)} \tag{5.6}$$

PROBLEMS AND NOTES

5.1 Construct for N = 6 a probability sampling of fixed size n = 3 such that it differs from simple random sampling but has the same probabilities of inclusion $\pi_i = 1/2$, $\pi_{ij} = 1/5$, $1 \le i \ne j \le 6$.

5.2 Carry out a simple random sample of size n = 40 from the population of digits in Table 6.1 (N = 1000) without numbering the elements. (*Hint:* We draw couples of numbers, the first indicating the row and the second the position in the row.)

5.3 Let N_0 be the nearest power of 10 higher than N. For example, if N = 3621, then N_0 = 10,000. What is the average number of numbers drawn randomly in the range $(1,N_0)$ before we obtain m numbers smaller than or equal to N? (*Answer*: mN_0/N)

5.4 Establish the expected number of independent draws of one integer from the range (1,N) before we get n *distinct* integers.

$$\left(\textit{Answer:}\ 1 + \frac{N}{N-1} + \cdots + \frac{N}{N-n+1} \simeq N \ln \frac{N + 1/2}{N - n + 1/2}\right)$$

5.5 Establish the expected number of independent draws from the range $(1,N_0)$, N_0 defined in Problem 5.3, before we obtain n distinct numbers smaller or equal to N.

$$\left(\textit{Answer:} \quad \frac{N_0}{N} + \frac{N_0}{N - 1} + \cdots + \frac{N_0}{N - n + 1} \simeq N_0 \ln \frac{N + 1/2}{N - n + 1/2} \right)$$

6

POISSON SAMPLING

In Poisson sampling the indicators of inclusion I_i are *independent* random variables. For any sequence of numbers $p_1, \ldots, p_N$, $0 \le p_i \le 1$, the corresponding Poisson sampling is defined by

$$P(s) = \prod_{i \in s} p_i \prod_{i \in r} (1 - p_i) \tag{6.1}$$

where $r = S - s$.

The main role of Poisson sampling is to help to define and analyze other sampling methods such as rejective sampling. These methods are obtained as conditional Poisson sampling. By conditioning we may remove the basic practical disadvantage of Poisson sampling--variability of the sample size. As we have seen in Theorem 3.3, every conditional Poisson sampling maximizes the entropy in the class of designs having the same π_i values and the same carrier.

THEOREM 6.1 In Poisson sampling

$$E(\text{size } s) = \sum_{i=1}^{N} p_i \tag{6.2}$$

$$\text{Var}(\text{size } s) = \sum_{i=1}^{N} p_i(1 - p_i) \tag{6.3}$$

$$\pi_i = p_i \qquad 1 \le i \le N \tag{6.4}$$

$$\pi_{ij} = p_i p_j \qquad 1 \le i \ne j \le N \tag{6.5}$$

Proof. We use the relations size $s = \Sigma_{i=1}^{N} I_i$, $\pi_i = EI_i$, $\pi_{ij} = EI_iI_j$, where the I_i are the indicators of inclusion introduced in Chapter 3. Then we make use of the fact that the indicators are independent under Poisson sampling and $EI_i = p_i$ and $\text{Var } I_i = p_i(1 - p_i)$.

If we wish the π_i to be proportionate to some numbers x_i, $\pi_i = \lambda x_i$ and the *expected* sample size to be n, then the condition $\Sigma_{i=1}^{N} \pi_i = \lambda \Sigma_{i=1}^{N} x_i = \lambda X = n$ provides $\lambda = n/X$, so that

$$p_i = \pi_i = \frac{nx_i}{X} \qquad 1 \le i \le N \tag{6.6}$$

Practically, Poisson sampling may be performed as follows. We decide that the p_i terms should be proportionate to some integral numbers x_i and that the expected sample size should be n. Then we associate every unit i with a number ξ_i randomly selected in the range $(1,X/n)$, where $X = \Sigma_{i=1}^{N} x_i$. If $\xi_i \le x_i$, we include the unit in the sample; if $\xi_i > x_i$, we do not. We may achieve--by multiplying the x_i terms by a power of 10 and by rounding off to integers--that the maximum x_i lies between 100 and 9999, which is a comfortable case. If the x_i terms attain only a few distinct values, it may be more expedient to multiply these values by $\lambda = 1000\ n/X$. Then ξ_i is selected in the range (1,1000) and is compared with λx_i, which may be found from an accessory table prepared in advance showing the correspondence between x_i and λx_i. If $\xi_i > \lambda \max x_i$, then we are sure that the unit will not be selected.

Example 6.1 In Table 6.1 there is a population of 1000 digits. As a first exercise, let us draw a sample of the expected size n = 100, and with p_i terms proportionate to x_i terms, where x_i denotes the digit in the ith place in Table 6.1. The total X equals

$$X = \sum_{i=1}^{1000} x_i = 4565$$

Thus $X/n = 4565/100 = 45.65$, and

$$100\,\frac{X}{n} = 4565$$

TABLE 6.1 A population of 1000 digits

6 3 6 4 2	2 9 8 2 7	3 2 7 2 3	6 1 9 3 7	2 1 9 2 8	4 8 5 7 2	1 8 2 7 2	7 2 4 6 9
3 1 8 7 4	7 9 1 2 3	7 2 1 8 5	4 7 1 9 8	6 2 5 3 9	5 4 1 7 6	3 2 4 7 9	2 8 7 3 2
4 2 6 3 8	1 5 2 2 5	8 7 2 3 9	6 1 3 2 5	1 7 2 9 3	4 1 3 8 2	3 1 4 6 3	1 8 2 4 5
9 4 5 3 1	2 6 4 9 8	3 2 1 6 9	4 2 6 5 2	8 5 3 1 8	3 5 2 8 2	3 7 4 2 5	2 4 6 2 7
3 6 2 1 9	8 4 6 7 3	2 6 4 9 1	4 7 3 5 2	9 2 9 1 8	6 3 4 5 2	5 9 3 2 5	7 8 8 2 3
2 7 3 5 6	3 9 4 3 3	7 4 2 8 1	6 7 3 4 9	5 3 2 8 4	2 9 8 2 3	4 7 3 2 5	8 4 3 1 6
6 3 2 7 9	4 5 5 4 1	8 5 4 5 2	9 3 9 2 1	4 8 3 1 5	3 8 2 6 3	9 4 2 3 7	6 3 1 2 4
8 2 6 2 4	9 3 2 7 4	5 8 2 6 4	3 7 1 9 7	8 6 2 3 4	3 5 4 5 6	8 2 4 3 1	7 2 2 4 6
7 2 3 8 9	4 3 9 5 1	2 6 7 3 4	8 7 2 1 6	9 3 4 3 1	2 6 6 7 8	4 6 7 1 3	6 5 1 9 8
3 2 3 4 9	2 4 8 5 1	5 6 2 8 1	7 6 3 4 2	2 5 4 2 7	3 6 5 4 1	3 2 4 5 6	8 9 2 1 2
9 4 8 1 2	5 9 2 8 7	1 6 7 7 2	4 4 3 3 8	2 4 3 9 6	3 9 7 2 6	5 1 6 9 8	3 5 4 2 6
6 3 2 1 3	8 4 3 1 6	2 4 6 9 8	7 5 2 6 3	4 6 9 8 3	2 1 4 5 5	4 2 1 6 3	5 4 9 2 1
4 3 3 5 8	9 1 4 4 6	9 9 4 5 2	3 7 8 5 6	2 4 7 6 5	4 9 2 1 3	8 6 5 1 4	2 3 8 9 6
5 9 1 2 6	7 3 4 4 8	2 6 3 7 3	8 2 1 3 8	2 9 4 3 2	7 4 5 8 2	4 3 2 1 3	7 2 4 3 1
6 3 2 1 6	2 4 3 9 1	4 6 8 9 3	5 6 3 2 1	9 4 3 6 5	8 2 1 6 5	4 5 6 1 7	2 3 6 5 8
4 2 6 3 1	8 5 4 3 2	9 1 6 6 4	2 3 4 8 2	3 1 6 5 5	7 2 4 6 3	2 7 1 6 4	9 3 2 2 2
5 2 4 6 8	1 2 4 9 3	6 5 4 3 8	6 5 6 2 3	7 5 4 1 2	6 1 4 4 8	6 2 3 4 1	9 8 7 3 5
7 3 4 2 6	5 4 8 5 7	6 3 2 1 5	3 7 8 2 6	4 1 7 5 3	4 7 7 5 3	9 2 6 5 3	4 8 7 3 4
6 5 7 2 3	8 4 5 2 5	7 6 3 7 4	6 2 5 7 8	1 4 3 5 2	4 5 6 5 3	7 2 3 7 3	4 6 5 7 5
2 5 3 8 5	7 4 1 5 9	3 6 4 5 2	9 7 2 4 3	5 4 2 6 7	3 9 5 2 6	8 1 4 8 2	5 6 4 3 2
3 1 7 4 5	2 6 4 3 3	9 5 7 2 3	4 3 7 8 7	2 6 3 5 1	8 5 4 2 5	7 3 4 3 6	8 5 2 4 7
5 4 3 1 2	3 7 2 6 3	4 5 6 5 4	2 7 4 3 2	7 5 2 5 3	4 8 2 5 6	1 9 6 5 7	2 4 6 5 8
7 1 5 2 4	3 6 5 7 5	4 2 8 9 3	1 6 5 7 2	8 5 2 3 4	6 1 5 2 7	4 2 5 8 7	3 1 4 3 5
4 3 5 1 8	2 7 6 3 5	7 4 2 1 6	5 4 2 6 9	8 5 4 6 3	2 5 4 7 3	6 7 5 2 1	9 8 2 4 3
5 8 7 2 3	4 7 3 4 2	6 5 3 7 1	9 4 2 6 5	7 2 5 4 6	8 6 3 7 7	5 5 4 3 2	7 8 5 2 1

Total = 4565

Note: The entries are denoted as x_i terms in the text.

TABLE 6.2 A Poisson sample of expected size $n = 100$ from the population of Table 6.1, with p_i terms proportionate to the entries x_i

<table>
<tr><td>6
5
9</td><td>5
4</td><td>7
7 9</td><td>9
2</td><td>8
8</td><td>8
5
3 4</td><td>2</td><td>7
2
5
7
8 3</td></tr>
<tr><td></td><td>3
5 5
4</td><td>8 5
6 4
5 6</td><td>7
6</td><td>8
8
2</td><td>3
1</td><td>3</td><td>6
9</td></tr>
<tr><td>3</td><td>8
9
4</td><td>3</td><td>4
3
6</td><td></td><td>8
5</td><td>6 9
6</td><td>7
6</td></tr>
<tr><td>6
5</td><td></td><td>6
6 7
6</td><td>2
6
7 8</td><td>6 5
3
6 7</td><td>6
3
6</td><td></td><td>2</td></tr>
<tr><td>5 2
7</td><td>2
7 4</td><td>9
6 5</td><td>4
5
6</td><td>3
7 5
8 4</td><td>8
7</td><td>5
6</td><td></td></tr>
</table>

For each unit we select a random integer ξ from the range (1,4565) and compare it with $100x_i$. If $\xi \le 100x_i$, the unit is selected; if $\xi > 100x_i$, the unit is not selected. For example, for the first unit in the first row we had $100x_i = 600$ and $\xi = 261$, so that this unit has been selected; on the other hand, for the second unit in the same row, we had $100x_i = 300$ and $\xi = 3262$, so that this unit was not included in the sample. One sample obtained by this procedure is given in Table 6.2, where the selected digits are situated in the same place as in the population.

The expected frequency of a digit k = 1, 2, ..., 9 in the sample equals

$$f_k = F_k \frac{k100}{4565}$$

where F_k is the frequency in the population. The values of F_k and f_k for the sample of Table 6.2 are given in Table 6.3.

Example 6.2 In Table 6.4 we have a sample of expected size n = 100 from the population of Table 6.1 with p_i terms proportionate to $x_i^{0.6}$.

TABLE 6.3 Actual and expected sample frequencies in the sample of Table 6.2

k	F_k	kF_k	f_k	Actual sample frequencies
1	83	83	1.8	1
2	163	326	7.1	7
3	146	438	9.6	11
4	135	540	11.8	9
5	119	595	13.0	16
6	109	654	14.3	21
7	96	672	14.7	13
8	84	672	14.7	12
9	65	585	12.8	7
Totals:	1000	4565	99.8	97

TABLE 6.4 A Poisson sample of expected size n = 100 from the population of Table 6.1 with the p_i terms proportionate to $x_i^{0.6}$

4 4 8 4	2 7 5 8	8 6	3 4	5 8 8 9	5 7 8 5	1 5	9 2
7 3 9 9	2 4 5		4 9 1 7 2 6	2 4	2 7	3 1	6 7
6 6	4 9 4	6 7 9	3 5	9 8 4	5	6 4	9
2 1 2 7	5 3 3	7 6 4	4 8		7	9 3 4 8	2 8 6 6 3
5	3 3 4 7 4	8	8 6 7 2 9	8 2	8	6 6 5	

TABLE 6.5 Powers of digits of Table 6.1

k	$k^{0.6}$	F_k	$k^{0.6}F_k$
1	1.000	83	83
2	1.519	163	247.63
3	1.933	146	282.24
4	2.297	135	310.14
5	2.626	119	312.56
6	2.930	109	319.38
7	3.214	96	308.65
8	3.482	84	292.49
9	3.737	65	242.92
		Total:	2398.91

We proceed as follows: First we compute the powers $k^{0.6}$ for $k = 1, 2, \ldots, 9$ and their population total, which yields Table 6.5. The total is rounded off to 2399. Now to speed up selection of random numbers from the range (1,2399) we identify 0 with 3 and 6, 1 with 4 and 7, and 2 with 5 and 8. Random numbers ξ_i are compared with $100x_i^{0.6}$. For example, for the first unit in the first row in Table 6.1 we have $x_i = 6$ and $100x_i^{0.6} = 293$, as is seen from Table 6.5. The selected random number is $\xi = 7326$, which is read as $\xi = 1326$. Since $\xi > 100x_i^{0.6}$, we do not include the unit in the sample. As for the fourth unit in the first row of Table 6.1, we have $100x_i^{0.6} = 230$ and $\xi = 6\ 192 = 0192$; since $192 < 230$, the unit is included in the sample. The expected and actual sample frequencies are given in Table 6.6.

EDGEWORTH EXPANSION Let us have a random variable y possessing a discrete or continuous density g(y) with expectation a, variance σ^2, and higher central moments $\mu_3, \mu_4, \ldots$. Put

$$x = \frac{y - a}{\sigma} \qquad \phi(x) = (2\pi)^{-1/2} \exp\left(-\frac{1}{2}x^2\right) \tag{6.7}$$

TABLE 6.6 Expected and actual sample frequencies for the sample of Table 6.4

k	f_k	Actual sample frequencies
1	3.5	4
2	10.3	11
3	11.8	10
4	12.9	16
5	13.0	11
6	13.3	13
7	12.9	12
8	12.2	14
9	10.1	11
Totals:	100.0	102

and

$$f_m(x) = \frac{1}{\sigma}\,\phi(x)\left[1 + \sum_{j=1}^{m} P_j(x)\right] \tag{6.8}$$

where $P_j(x)$ are polynomials derived by Edgeworth and to be found in Cramér (1946, Sec. 17.7). $P_j(x)$ is a linear combination of Hermite polynomials of order $j + 2$, $j + 4$, ..., $3j$. For example,

$$P_1(x) = \frac{1}{3!}\,(x^3 - 3x)\,\frac{\mu_3}{\sigma^3}$$

$$P_2(x) = \frac{1}{4!}\left(\frac{\mu_4}{\sigma^4} - 3\right)(x^4 - 6x^2 + 3) + \frac{10}{6!}\left(\frac{\mu_3}{\sigma^3}\right)^2 (x^6 - 15x^4 + 45x^2 - 15) \tag{6.9}$$

We shall call $f_m(x)$ of (6.8) the m-term Edgeworth expansion of the density $g(y)$.

We shall apply the Edgeworth expansion to the random variable K = size s in Poisson sampling. We have $K = \Sigma_{i=1}^{N} I_i$, where I_i are independent indicators with $EI_i = p_i$, $1 \le i \le N$. Thus

$$\begin{aligned} EK &= \sum_{i=1}^{N} p_i = n \\ \operatorname{Var} K &= \sum_{i=1}^{N} p_i(1 - p_i) = d \\ \mu_3 &= \sum_{i=1}^{N} p_i(1 - p_i)(1 - 2p_i) \\ \mu_4 &= \sum_{i=1}^{N} p_i(1 - p_i)[1 - 6p_i(1 - p_i)] + 3d^2 \end{aligned} \tag{6.10}$$

and so on. The standardized random variable will be

$$x = \frac{k - n}{\sqrt{d}} \tag{6.11}$$

SYMBOLS $O(\cdot)$ AND $o(\cdot)$ The symbol $O(\varepsilon)$ will denote a number such that $O(\varepsilon)/\varepsilon$ remains bounded if $\varepsilon \to 0$. The symbol $o(\varepsilon)$ will denote a number such that $[o(\varepsilon)/\varepsilon] \to 0$ if $\varepsilon \to 0$.

For example, $(n + 5)/n^2 = O(1/n) = o(n^{-1/2})$.

THEOREM 6.2 (V. V. Petrov) Under Poisson sampling we have

$$|P(\text{size } s = k) - f_m(x)| = o[d^{-(m+1)/2}] \tag{6.12}$$

where $f_m(x)$ is the m-term Edgeworth expansion (6.8), d = Var size s, and x is related to k by (6.11), uniformly in k.

Proof. V. V. Petrov (1962) proved the theorem for the case $d = O(n)$, which is the most important case. However, the theorem is valid for any rate of convergence of $d \to \infty$. The result is obtained from the formula

$$P(\text{size } s = k) = \frac{1}{2\pi} \int_{-\pi}^{\pi} e^{-itk} \psi(t)\, dt$$

where $\psi(t)$ is the characteristic function of K = size s,

$$\psi(t) = \prod_{j=1}^{N} (1 - p_j + p_j e^{it})$$

We can see that

$$|1 - p_j + p_j e^{it}|^2 = 1 - 2p_j(1 - p_j)(1 - \cos t)$$

$$\leq \frac{\pi^2 - p_j(1 - p_j)t^2}{\pi^2} \leq e^{-p_j(1-p_j)t^2/\pi^2}$$

and, consequently,

$$|\psi(t)|^2 \leq e^{-dt^2/\pi^2}$$

Thus outside the origin, $\psi(t)$ approaches zero faster than any power of 1/d as $d \to \infty$. This takes care of the tails of $\psi(t)$. In the vicinity of the origin we expand $\psi(t)$ as follows:

$$\psi(t) = e^{itn-(1/2)t^2 d}\left[1 + \sum_{j=1}^{m} Q_j(it) + R(t)\right]$$

where $Q_j(it)$ are polynomials arranged so that the terms of $Q_j(iu/d^{1/2})$ are of order $O(d^{-j/2})$. The result then easily follows.

Remark 6.1 From (6.8), (6.9), and (6.10) it follows that a two-term Edgeworth expansion for P(size s = k) equals

$$f_2(x) = \frac{1}{d^{1/2}}\,\phi(x)\left[1 + \frac{1 - 2\bar{\bar{p}}}{6d^{1/2}}(x^3 - 3x) + \frac{1}{4!}\,\frac{1 - 6\overline{\overline{p(1 - p)}}}{d}(x^4 - 6x^2 + 3) + \frac{10}{6!}\,\frac{(1 - 2\bar{\bar{p}})^2}{d}(x^6 - 15x^4 + 45x^2 - 15)\right] \quad (6.13)$$

where

$$\bar{\bar{p}} = \frac{\sum_{i=1}^{N} p_i^2(1 - p_i)}{\sum_{i=1}^{N} p_i(1 - p_i)} \qquad \overline{\overline{p(1 - p)}} = \frac{\sum_{i=1}^{N} p_i^2(1 - p_i)^2}{\sum_{i=1}^{N} p_i(1 - p_i)} \quad (6.14)$$

TABLE 6.7 Binomial probabilities for N = 100 and p = 0.1 compared with the normal approximation and with one-term Edgeworth expansion

k	Binomial probabilities	Normal approximation	Edgeworth approximation
0	.0000	.0005	-.0001
1	.0003	.0015	.0003
2	.0016	.0039	.0019
3	.0059	.0089	.0065
4	.0159	.0183	.0164
5	.0339	.0334	.0337
6	.0596	.0549	.0586
7	.0889	.0807	.0878
8	.1148	.1062	.1145
9	.1304	.1253	.1311
10	.1319	.1324	.1330
11	.1199	.1253	.1204
12	.0988	.1062	.0984
13	.0743	.0807	.0735
14	.0513	.0549	.0507
15	.0327	.0334	.0326
16	.0193	.0183	.0196
17	.0106	.0089	.0109
18	.0054	.0039	.0056
19	.0026	.0015	.0026
20	.0012	.0005	.0011
21	.0005	.0002	.0004
22	.0002	.0000	.0002
23	.0001	.0000	.0000

Remark 6.2 The remainder in using $f_2(x)$ is $o(d^{-3/2})$, which is smaller than the effect of discretizing the normal distribution, i.e., replacing $d^{-1/2}\phi(x)$ by

$$\int_{x-(1/2)d^{-1/2}}^{x+(1/2)d^{-1/2}} \phi(y)\, dy$$

The exactness of the Edgeworth expansion in approximating P(size s = k) is thus amazing.

Example 6.3 Table 6.7 illustrates the improvement over the normal approximation achieved by using the one-term Edgeworth expansion for binomial distributions with $n = 100$ and $p = 0,1$. The first two columns of the table are taken from Hald (1952), where the normal approximation is not based on normal density, but on normal integrals over unit intervals. Note that the above binomial distribution describes the distribution of size s in Poisson sampling with $N = 100$ and $\pi_i = 0.1$, $1 \le i \le 100$.

PROBLEMS AND NOTES

6.1 Prove that the distribution of the random variable K = size s in arbitrary Poisson sampling is strongly unimodal in the sense that

$$P_k^2 \ge P_{k-1}P_{k+1} \qquad k = 1, \ldots, N - 1$$

where $P_k = P(K = k)$. (Samuels, 1965; or you may use induction from N to N + 1.)

6.2 By means of Problem 6.1 show that the distribution of K = size s is unimodal in Poisson sampling.

6.3 From the population of Table 6.1 carry out a Poisson sample of expected size $n = 20$ with probabilities proportionate to $\sqrt{x_i}$.

7

REJECTIVE SAMPLING

Rejective sampling may be defined either as conditional Poisson sampling or as conditional sampling with replacement. We shall first describe it as conditional Poisson sampling given size s = n. If the parameters of the corresponding Poisson sampling are $p_1, \ldots, p_N$, the formula for rejective sampling obtained from it by fixing size s = n will be

$$P(s) = \begin{cases} c \prod_{i \in s} p_i \prod_{i \in r} (1 - p_i) & \text{if size } s = n \\ 0 & \text{otherwise} \end{cases} \tag{7.1}$$

where the constant c is chosen so as to achieve $\Sigma P(s) = 1$. Obviously $c = c(p_1, \ldots, p_N, n)$. The association of a rejective sampling with a Poisson sampling given by (7.1) is not unique. Any other set of probabilities $(p_1^*, \ldots, p_N^*)$ provides the same rejective sampling if

$$\frac{p_i}{1 - p_i} = \lambda \frac{p_i^*}{1 - p_i^*} \qquad 1 \le i \le N$$

for some λ.

We shall call the representation (7.1) *canonical* if n equals the expected sample size under the corresponding Poisson sampling, i.e., if

$$n = \sum_{i=1}^{N} p_i$$

Obviously there is exactly one canonical representation of rejective sampling by (7.1).

Another definition of rejective sampling is as follows:

$$P(s) = \begin{cases} c^* \prod_{i \in s} \alpha_i & \text{if size } s = n \\ 0 & \text{otherwise} \end{cases} \tag{7.2}$$

where

$$0 \le \alpha_1, \ldots, \alpha_N \le 1 \qquad \alpha_1 + \cdots + \alpha_N = 1 \tag{7.3}$$

It is easy to see that definitions (7.1) and (7.2) are equivalent if we put

$$\alpha_i = \lambda \frac{p_i}{1 - p_i} \qquad 1 \le i \le N \tag{7.4}$$

where the value of λ follows from (7.3).

Formula (7.2) corresponds to the definition of rejective sampling as conditional sampling with replacement. *Sampling with replacement* consists of n subsequent draws of units one by one in such a way that

1. The probability of selection of unit i in every draw is α_i, $1 \le i \le N$.

2. The draws are independent.

If we add the condition that all n selected units should be *distinct*, the respective conditional sampling with replacement becomes rejective sampling and P(s) will be given by (7.2). Thus rejective sampling may be regarded as sampling with replacements modified by

3. If we draw a unit selected at some previous draw, we reject all the partially built up samples and start anew until we select a sequence of n distinct units without a repetition occurring in between.

DRAWING PROBABILITIES The probability α_i of selecting the unit i in individual draws will be called drawing probability. This has to be strictly distinguished from the inclusion probabilities π_i. We have $\Sigma_{i=1}^N \alpha_i = 1$, whereas for fixed size sampling we have $\Sigma_{i=1}^N \pi_i = n$.

The first method of realization of rejective sampling--as conditional Poisson sampling--is impractical, since the probability of rejecting a sample for size $s \neq n$ will be usually too large.

THEOREM 7.1 In Poisson sampling with probabilities of inclusion $p_1, \ldots, p_N$, $p_1 + \cdots + p_N = n$,

$$P(\text{size } s = n) \sim \left[2\pi \sum_{i=1}^{N} p_i(1 - p_i)\right]^{-1/2}$$

$$\text{if } \Sigma\, p_i(1 - p_i) \to \infty \qquad (7.5)$$

holds, where ~ means that the ratio of both sides goes to 1 as $\Sigma_{i=1}^N p_i(1 - p_i)$ tends to infinity.

Proof. Expressing size $s = \Sigma_{i=1}^N I_i$ and applying Theorem 6.2, we obtain

$$P(\text{size } s = k) \sim \left[2\pi \sum_{i=1}^{N} p_i(1 - p_i)\right]^{-1/2} \exp\left[-\frac{1}{2}\frac{(k - n)^2}{\sum_{i=1}^N p_i(1 - p_i)}\right]$$

which yields (7.5) for k = n. Q.E.D.

Remark 7.1 The result of Theorem 7.1 could be equivalently expressed as follows: For every $\varepsilon > 0$ there exists a d_ε such that for any Poisson sampling with probabilities of inclusion $p_1, \ldots, p_N$ such that

$$\sum_{i=1}^{N} p_i(1 - p_i) > d_\varepsilon$$

and $p_1 + \cdots + p_N = n$ = integer, the conclusion

$$\left| P(\text{size } s = n)\left[2\pi \sum_{i=1}^{N} p_i(1 - p_i)\right]^{1/2} - 1 \right| < \varepsilon$$

holds.

We also could express the result of Theorem 7.1 in terms of sequences. Then, considering a sequence of populations of sizes N_ν, $1 \le \nu < \infty$, and a corresponding sequence of Poisson samples with probabilities of inclusion $(p_{\nu 1}, \ldots, p_{\nu N_\nu})$, $1 \le \nu < \infty$, we would say that

$$\lim_{\nu \to \infty} \sum_{i=1}^{N_\nu} p_{\nu i}(1 - p_{\nu i}) = \infty$$

entails

$$\lim_{\nu \to \infty} P_\nu(\text{size } s_\nu = n_\nu) \left[2\pi \sum_{i=1}^{N_\nu} p_{\nu i}(1 - p_{\nu i}) \right]^{1/2} = 1$$

provided that $\Sigma_{i=1}^{N_\nu} p_{\nu i} = n_\nu$ = an integer, $1 \le \nu < \infty$.

Still another way of expressing our result would be through the symbol $o(\cdot)$ denoting "smaller order than." Then we would write

$$P(\text{size } s = n) = \left[2\pi \sum_{i=1}^{N} p_i(1 - p_i) \right]^{-1/2} [1 + o(1)]$$

$$\text{as } \sum_{i=1}^{N} p_i(1 - p_i) = d \to \infty$$

where $o(1)$ denotes a number such that $o(1) \to 0$ as $d \to \infty$.

The last but one way of expressing the results is probably most instructive, but the present author finds the repeated use of the sequence label ν so tedious that he avoids it. However, the reader is urged to understand correctly the statement of Theorem 7.1 and of similarly formulated theorems below.

Remark 7.2 From (7.5) we obtain a simple approximate upper bound for P(size s = n):

$$P(\text{size } s = n) < \frac{1}{\sqrt{6.28 \lambda n}} \tag{7.6}$$

where λ is some number satisfying $\lambda \le (1 - \pi_i)$, $1 \le i \le N$, and where we have put $2\pi \doteq 6.28$. For usual sizes of n and $\lambda \ge 1/2$ we obtain from (7.6) an idea of how small P(size s = n) is and how often we would have to repeat the sampling.

Example 7.1 If n = 100, λ = 0.9, we obtain

$$P(\text{size } s = n) < \frac{1}{\sqrt{6.28 \times 0.9 \times 100}} = \frac{1}{\sqrt{565.2}} \doteq 0.042$$

That means that the sampling would have to be repeated in the average at least 23 times.

If rejective sampling is realized as conditional sampling with replacement, the probability of rejection will be a little bit more involved.

THEOREM 7.2 Let A denote the event that all the units drawn at n subsequent independent draws with probabilities $\alpha_1, \ldots, \alpha_N$ will be distinct. Then

$$P(A) \sim n! \left(\sum_{i=1}^{N} \frac{p_i}{1 - p_i} \right)^{-n} \left[\prod_{i=1}^{N} (1 - p_i) \right]^{-1} \left[2\pi \sum_{i=1}^{N} p_i(1 - p_i) \right]^{-1/2} \tag{7.7}$$

as $\Sigma_{i=1}^{N} p_i(1 - p_i) \to \infty$. The numbers α_i and p_i are related by (7.4) with the side condition $p_1 + \cdots + p_N = n$.

Proof. The probability of a sequence of distinct units consisting of a sample s obviously equals $\Pi_{i\in s} \alpha_i$. To each sample, however, there correspond n! such sequences all having the same probability. Thus the probability of obtaining a sample s equals

$$P(s) = \frac{n! \, \Pi_{i\in s} \alpha_i}{P(A)} \tag{7.8}$$

Comparing (7.8) with (7.2) we can see that

$$P(A) = \frac{n!}{c^*} \tag{7.9}$$

On the other hand, comparing (7.2) and (7.1) and substituting

$$\alpha_i = \frac{p_i/(1 - p_i)}{\sum_{j=1}^{N} p_j/(1 - p_j)} \tag{7.10}$$

where $p_1 + \cdots + p_N = n$, we can see that

$$\frac{1}{c^*} = \frac{[\sum_{i=1}^N p_i/(1 - p_i)]^{-n}[\prod_{i=1}^N (1 - p_i)]^{-1}}{c} \tag{7.11}$$

where c is from (7.1). However,

$$\frac{1}{c} = P(\text{size } s = n) \sim \frac{1}{\sqrt{2\pi \Sigma_{i=1}^N p_i(1 - p_i)}} \tag{7.12}$$

Putting (7.9) through (7.12) together, we obtain (7.7). Q.E.D.

Remark 7.3 If the sum $\Sigma_{i=1}^N p_i^2$ is not large, we may approximate

$$\sum_{i=1}^N \frac{p_i}{1 - p_i} \sim n\left(1 + \frac{1}{n}\sum_{i=1}^N p_i^2\right) \sim n \exp\left(\frac{1}{n}\sum_{i=1}^N p_i^2\right)$$

$$\prod_{i=1}^N (1 - p_i) \sim \exp\left(-n - \frac{1}{2}\sum_{i=1}^N p_i^2\right)$$

$$n! \sim (2\pi)^{1/2} n^{n+(1/2)} e^{-n}$$

$$\left[2\pi \sum_{i=1}^N p_i(1 - p_i)\right]^{-1/2} \sim (2\pi)^{-1/2} n^{-1/2} \exp\left(\frac{1}{2}\frac{1}{n}\sum_{i=1}^N p_i^2\right) \tag{7.13}$$

which yields

$$P(A) \sim \exp\left(-\frac{1}{2}\frac{n - 1}{n}\sum_{i=1}^N p_i^2\right) \tag{7.14}$$

Introducing the coefficient of variation of the p_i terms, $1 + V_p^2 = N \Sigma_{i=1}^N p_i^2/n^2$, we can write

$$P(A) \sim \exp\left[-\frac{1}{2}\frac{n(n - 1)}{N}(1 + V_p^2)\right] \tag{7.15}$$

Example 7.2 Consider n = 50, N = 5000, and $V_p^2 = 0.2$. Then

$$P(A) \sim \exp\left(-\frac{1}{2}\frac{50 \times 49}{5000} 1.2\right) \doteq 0.745$$

Now we shall investigate the most important aspect of rejective sampling--expressing the π_i in terms of p_i and α_i and π_{ij} in terms of π_i, $1 \leq i \leq N$.

THEOREM 7.3 Consider rejective sampling given by (7.1) where $p_1 + \cdots + p_N = n$. Then the probabilities of inclusion π_i equal

$$\pi_i = p_i\left[1 - \frac{(\bar{\bar{p}} - p_i)(1 - p_i)}{d} + o(d^{-1})\right] \qquad 1 \le i \le N \tag{7.16}$$

where

$$d = \sum_{i=1}^{N} p_i(1 - p_i) \tag{7.17}$$

and

$$\bar{\bar{p}} = \frac{\sum_{i=1}^{N} p_i^2(1 - p_i)}{d} \tag{7.18}$$

The term $o(d^{-1})$ satisfies $d\, o(d^{-1}) \to 0$ as $d \to \infty$, uniformly in i, $1 \le i \le N$.

Proof. Consider Poisson sampling corresponding to $p_1, \ldots, p_N$ and note that

$$\pi_i = P(I_i = 1|\text{size } s = n) \tag{7.19}$$

where $P(\cdot)$ refers to the Poisson sampling. Utilizing the Bayes formula, we also have

$$\pi_i = P(I_i = 1)\,\frac{P(\text{size } s = n|I_i = 1)}{P(\text{size } s = n)} \tag{7.20}$$

Now, according to Theorem 6.2 and (6.13), if we take the two-term series (m = 2),

$$P(\text{size } s = n) = \frac{1}{\sqrt{2\pi d}}\,[1 + cd^{-1} + o(d^{-1})] \tag{7.21}$$

where

$$c = \frac{1}{8}\,(1 - 6\overline{\overline{p(1 - p)}}) - \frac{5}{24}\,(1 - \bar{\bar{p}})^2$$

with $\overline{\overline{p(1 - p)}}$ and $\bar{\bar{p}}$ defined by (6.14).

Now conditionally on $I_i = 1$ the random variable size s may be expressed as

$$\text{Size s} = \sum_{j \neq i}^{N} I_j + 1$$

where the I_j terms are independent. Thus

$$E(\text{size s} | I_i = 1) = \sum_{j \neq i}^{N} p_j + 1 = n - p_i + 1$$

$$\text{Var}(\text{size s} | I_i = 1) = \sum_{j \neq i}^{N} p_j(1 - p_j) = d - p_i(1 - p_i)$$

The third central moment of I_i equals

$$(1 - p_i)^3 p_i + (0 - p_i)^3(1 - p_i) = p_i(1 - p_i)(1 - 2p_i)$$

Thus the conditional third central moment of size s equals

$$\begin{aligned}\mu_3 &= E[(\text{size s} - n - 1 + p_i)^3 | I_i = 1] \\ &= \sum_{j \neq i}^{N} p_j(1 - p_j)(1 - 2p_j) \\ &= d(1 - 2\bar{\bar{p}}) - p_i(1 - p_i)(1 - 2p_i)\end{aligned}$$

and the conditional fourth moment equals

$$\begin{aligned}\mu_4 &= \sum_{j \neq i}^{N} p_j(1 - p_j)[1 - 6p_j(1 - p_j)] + 3[d - p_i(1 - p_i)]^2 \\ &= d[1 - 6\overline{\overline{p(1 - p)}}] - p_i(1 - p_i)[1 - 6p_i(1 - p_i)] \\ &\quad + 3[d - p_i(1 - p_i)]^2\end{aligned}$$

Applying Theorem 6.2 with m = 2, we obtain

$$
\begin{aligned}
&P(\text{size } s = n \mid I_i = 1) \\
&= [2\pi(d - p_i(1 - p_i))]^{-1/2} \exp\left[-\frac{1}{2}\frac{(n - n + p_i - 1)^2}{d - p_i(1 - p_i)}\right] \\
&\times \left\langle 1 + \frac{d(1 - 2\bar{\bar{p}}) - p_i(1 - p_i)(1 - 2p_i)}{6[d - p_i(1 - p_i)]^{3/2}} \right. \\
&\times \left\{ x^3 - 3\frac{n - n + p_i - 1}{[d - p_i(1 - p_i)]^{1/2}} \right\} \\
&+ \frac{d[1 - 6\overline{\overline{p(1 - p)}}] - p_i(1 - p_i)[1 - 6p_i(1 - p_i)]}{4![d - p_i(1 - p_i)]^2} \\
&\times (x^4 - 6x^2 + 3) \\
&+ \frac{10[d(1 - 2\bar{\bar{p}}) - p_i(1 - p_i)(1 - 2p_i)]^2}{6![d - p_i(1 - p_i)]^3} \\
&\left. \times (x^6 - 15x^4 + 45x^2 - 15) + o(d^{-1}) \right\rangle
\end{aligned}
$$

where $x = (p_i - 1)[d - p_i(1 - p_i)]^{-1/2}$. Now we may simplify as follows:

$$
\begin{aligned}
P(\text{size } s = n \mid I_i = 1) &= (2\pi d)^{-1/2}\left[1 + \frac{1}{2}\frac{p_i(1 - p_i)}{d} - \frac{1}{2}\frac{(1 - p_i)^2}{d} \right. \\
&\qquad \left. - \frac{1}{2}\frac{(p_i - 1)(1 - 2\bar{\bar{p}})}{d} + \frac{c}{d} + o(d^{-1})\right] \\
&= (2\pi d)^{-1/2}\left[1 - \frac{(\bar{\bar{p}} - p_i)(1 - p_i)}{d} \right. \\
&\qquad \left. + \frac{c}{d} + o(d^{-1})\right] \qquad (7.22)
\end{aligned}
$$

Substituting (7.21), (7.22), and $P(I_i = 1) = p_i$ into (7.20), we obtain (7.16). Q.E.D.

THEOREM 7.4 In rejective sampling

$$\pi_{ij} = \pi_i \pi_j \left[1 - \frac{(1 - \pi_i)(1 - \pi_j)}{d} + o(d^{-1}) \right] \tag{7.23}$$

holds, where

$$d = \sum_{i=1}^{N} \pi_i (1 - \pi_i) \tag{7.24}$$

and $d\, o(d^{-1}) \to 0$ as $d \to \infty$ uniformly in i, j, $1 \leq i \neq j \leq N$.

Proof. Similarly as in the preceding proof, an application of Theorem 6.2 yields

$$\pi_{ij} = p_i p_j \left[1 + \frac{1}{2} \frac{p_i(1 - p_i) + p_j(1 - p_j)}{d} - \frac{1}{2} \frac{(2 - p_i - p_j)^2}{d} \right.$$

$$\left. - \frac{1}{2} \frac{(p_i + p_j - 2)(1 - 2\bar{\bar{p}})}{d} + o(d^{-1}) \right] \tag{7.25}$$

Now, in view of (7.16), we also have

$$\pi_{ij} = \pi_i \pi_j \left[1 - \frac{(1 - p_i)(1 - p_j)}{d} + o(d^{-1}) \right] \tag{7.26}$$

Finally, in view of (7.16), we may replace p_i by π_i and d of (7.17) by d of (7.24). Q.E.D.

Theorem 7.4 provides the following asymptotically valid approximation:

$$\pi_i \pi_j - \pi_{ij} \sim \frac{\pi_i(1 - \pi_i)\pi_j(1 - \pi_j)}{d} \qquad 1 \leq i \neq j \leq N \tag{7.27}$$

The approximation on the right-hand side of (7.27) has the form of a product $c_i c_j$ where $c_i = \pi_i(1 - \pi_i)/\sqrt{d}$. However, it does not satisfy the relations $\Sigma_{j\neq i}^{N} c_i c_j = \Sigma_{j\neq i}^{N} (\pi_i \pi_j - \pi_{ij}) = \pi_i(1 - \pi_i)$, $1 \leq i \leq N$. To achieve this we must put $c_i = \pi_i(1 - \lambda_i)/\sqrt{d_\lambda}$, where $d_\lambda = \Sigma_{i=1}^{N} \pi_i(1 - \lambda_i)$ and the λ_i terms are given implicitly by (3.24), and explicitly, but approximately, by (3.26) and (3.27). Then we have

$$\pi_i\pi_j - \pi_{ij} \sim c_ic_j = \frac{\pi_i(1-\lambda_i)\pi_j(1-\lambda_j)}{\sum_{k=1}^N \pi_k(1-\lambda_k)} \qquad 1 \le i \ne j \le N \tag{7.28}$$

This approximation is also asymptotically valid, since $(1 - \lambda_i) \sim (1 - \pi_i)$ holds uniformly in $1 \le i \le N$, as may be seen from (3.24). Approximation (7.28) will be called *tight*, indicating that $\Sigma_{j\ne i}^N c_ic_j = \pi_i(1 - \pi_i)$, $1 \le i \le N$.

Example 7.3 Let us test the fit of formula (7.28) on an example with $N = 10$, $n = 4$, and the π_i terms as follows:

$\pi_1 = .0901$ $\quad\pi_4 = .3263$ $\quad\pi_8 = .5580$
$\pi_2 = .1745$ $\quad\pi_5 = .3933$ $\quad\pi_9 = .6014$
$\pi_3 = .2533$ $\quad\pi_6 = .4542$ $\quad\pi_{10} = .6399$
$\pi_7 = .5091$

We have

$$\frac{\max \pi_i}{\min \pi_i} = \frac{.6399}{.0901} = 6.1$$

Table 7.1 shows exact values of $\pi_i\pi_j - \pi_{ij}$ in its top part, and the approximation based on formula (7.28) with

$$1 - \lambda_i = \frac{1 - \pi_i}{1 - \pi_i(1 - \pi_i)/\Sigma_{j=1}^N \pi_j(1 - \pi_j)} \qquad 1 \le i \le N \tag{7.28a}$$

in its bottom part. The fit is fairly good if we take into account that the formula is valid asymptotically, and we have $d < 3$ and large and very variable π_i terms. For larger d and for smaller or less variable π_i terms the fit would still improve.

THEOREM 7.5 The single-draw probabilities α_i producing given probabilities of inclusion π_i in rejective sampling (7.2) are equal to

$$\alpha_i = \lambda\left[(d - 1 + \bar{\bar{\pi}})\frac{\pi_i}{1 - \pi_i} + \pi_i + o(1)\right] \qquad 1 \le i \le N \tag{7.29}$$

TABLE 7.1 Testing the fit of Eq. (7.28)

i \ j	2	3	4	5	6	7	8	9	10
1	51	70	85	96	102	105	106	104	101
2		129	155	174	185	190	190	186	180
3			211	235	250	255	254	248	239
4				281	296	302	299	291	279
5					326	331	326	316	302
6						344	338	326	311
7							338	325	309
8								316	300
9									286

$10^4(\pi_i\pi_j - \pi_{ij})$ --exact values

i \ j	2	3	4	5	6	7	8	9	10
1	56	76	90	98	103	104	102	99	95
2		138	163	178	186	188	185	179	171
3			219	240	250	253	249	241	230
4				283	296	298	294	285	272
5					324	327	322	312	298
6						342	337	326	312
7							340	329	315
8								324	310
9									300

Approximation based on Eq. (7.28)

where d is given by (7.24) and

$$\bar{\bar{\pi}} = \frac{\sum_{i=1}^{N} \pi_i^2(1 - \pi_i)}{d}$$

Again d $o(d^{-1}) \to 0$ as $d \to \infty$, uniformly in i = 1, ..., N.

Proof. From (7.16) it follows that

$$p_i = \pi_i\left[1 + \frac{(\bar{\bar{\pi}} - \pi_i)(1 - \pi_i)}{d} + o(d^{-1})\right] \tag{7.30}$$

Combining (7.30) with (7.4) we can see that

$$\alpha_i = \lambda \frac{\pi_i[1 + (\bar{\bar{\pi}} - \pi_i)(1 - \pi_i)/d + o(d^{-1})]}{(1 - \pi_i)[1 - (\bar{\bar{\pi}} - \pi_i)\pi_i/d + o(d^{-1})]}$$

$$= \lambda \frac{\pi_i}{1 - \pi_i}\left[1 + \frac{\bar{\bar{\pi}} - \pi_i}{d} + o(d^{-1})\right]$$

$$= \lambda \frac{\pi_i}{1 - \pi_i}[d + \bar{\bar{\pi}} - 1 + (1 - \pi_i) + o(1)]$$

Q.E.D.

Remark 7.4 Theorem 7.5 provides the approximate formula

$$\alpha_i \sim \lambda\left[(d - 1 + \bar{\bar{\pi}})\frac{\pi_i}{1 - \pi_i} + \pi_i\right] \tag{7.31}$$

which shows that the α_i terms should be a weighted average of $\pi_i/(1 - \pi_i)$ and π_i, the relative weight of π_i tending to 0 as $d \to \infty$. Formula (7.31) works very well even for small d, even for n = 1.

Example 7.4 Let us try to find drawing probabilities α_i which would yield rejective sampling of size n = 4 from a population of N = 10 with inclusion probabilities $\pi_1, \ldots, \pi_{10}$ given in the second row of Table 7.2. Applying formula (7.31) we obtain the α_i terms that are presented in the last row of Table 7.2. Next, utilizing formula (7.2) we compute P(s) for all $\binom{10}{4} = 210$ samples of size n = 4 and then establish the inclusion probabilities, denoted as π_i^*, according to formula (3.3). The achieved probabilities π_i^* are given in the third row of Table 7.2. We can see that π_i^* approximate excellently the target probabilities π_i.

TABLE 7.2 Effectiveness of Eq. (7.31)

i	1	2	3	4	5	6	7	8	9	10
π_i	.0902	.1747	.2534	.3262	.3930	.4537	.5086	.5578	.6017	.6408
π_i^*	.0901	.1745	.2533	.3263	.3933	.4542	.5091	.5580	.6014	.6399
α_i	.0147	.0301	.0467	.0645	.0836	.1042	.1264	.1503	.1760	.2036

Remark 7.5 Example 7.4 is a side product of a numerical experiment in which we compare the values of $\pi_i\pi_j - \pi_{ij}$ for a rejective sampling (Table 7.1) and a successive sampling (Table 9.2), both having nearly the same first-order inclusion probabilities. We start with the successive sampling considered in Table 9.2, which results in inclusion probabilities given in the second row of Table 7.2, and then, utilizing formula (7.31), we construct rejective sampling, which yields inclusion probabilities given in the third row of Table 7.2. Both sets of inclusion probabilities do not coincide exactly, but are sufficiently close to each other. The comparison of the values for $\pi_i\pi_j - \pi_{ij}$ is commented upon in Example 9.3.

Remark 7.6 Formula (7.31) may be rewritten as follows:

$$\alpha_i \sim \lambda\pi_i \frac{1 - \pi_i/(d + \bar{\bar{\pi}})}{1 - \pi_i}$$

If the π_i terms are small, we can replace $d + \bar{\bar{\pi}}$ by n and obtain the following chain of approximations:

$$\alpha_i \doteq \lambda\pi_i \frac{1 - \pi_i/n}{1 - \pi_i} \doteq \frac{\lambda\pi_i}{1 - (n - 1)\pi_i/n} \doteq \lambda\pi_i\left(1 + \frac{n - 1}{n}\pi_i\right)$$

$$\doteq \frac{1}{n}\pi_i\left[1 + \frac{n - 1}{n}\left(\pi_i - \frac{1}{n}\sum_{j=1}^{N}\pi_j^2\right)\right] \tag{7.32}$$

where λ is a norming constant changing form case to case. Of course, for practical sampling it often suffices to put $\alpha_i = \lambda\pi_i(1 + \pi_i)$, or $\alpha_i = \lambda\pi_i/(1 - \pi_i)$, or even $\alpha_i = \pi_i/n$. If $\pi_i = nx_i/X$, the last line in (7.32) becomes

$$\alpha_i \doteq \frac{x_i}{X}\left[1 + \frac{(n - 1)x_i}{X} - (n - 1)\sum_{j=1}^{N}\left(\frac{x_j}{X}\right)^2\right] \tag{7.33}$$

On the basis of Chapter 15 and Hájek (1964), we can also approximately realize rejective sampling as a *corrected Poisson sampling*. We proceed as follows: Assume we want to obtain a rejective sample of size n and with π_i terms proportionate to some x_i terms. We select a Poisson sample s' in accordance with inclusion probabilities $p_i = nx_i/X$, $1 \le i \le N$. The achieved sample size, $n' = \text{size } s'$, will generally be different from n.

If $n' > n$, we regard s' as a population and remove from it $n' - n$ units by rejective sampling of size $n' - n$ and with probabilities π_i proportionate to $X - nx_i$.

If $n' < n$, we regard $S - s'$ as a population and draw from it a rejective sample of size $n - n'$ with probabilities π_i proportionate to x_i, and add the selected units to s'.

If $n' = n$, we accept s' without any change.

In all three cases we obtain as a final sample a subset of n units, which may be regarded as an approximation to the desired rejective sample.

Example 7.5 Let us correct a sample of Table 6.2 so as to obtain a rejective sample of size $n = 100$ with π_i terms proportionate to the x_i terms. The actual size of the Poisson sample equals $n' = 97 < n = 100$. So we have to select a rejective sample of size 3 from the remainder of the population with probabilities π_i proportionate to x_i. We shall realize this rejective sampling as conditional sampling with replacement with proper drawing probabilities α_i, $i \in S - s'$. Since the sample total is $x = \Sigma_{i \in s'} x_i = 540$, the total over the rest equals $4565 - 540 = 4025$ and the inclusion probabilities will equal $\pi_i = 3x_i/4025$. Since the π_i terms are very small for any x_i in the population, the drawing probabilities α_i may also be proportionate to the x_i terms, as is apparent from (7.31), so that we may put

$$\alpha_i = \frac{x_i}{4025} \qquad i \in S - s'$$

Sampling of one unit with these probabilities may be carried out without computing the successive total of the x_i terms by the following method due to Lahiri (1951): We draw one element by random sampling with uniform probabilities, then we draw a random integer ξ in the range (1,9); if $\xi \le x_i$, we accept the unit, otherwise we reject it. A random draw of one unit from the population of Table 6.1 may be further simplified by drawing randomly two numbers (α,β),

α indicating the row and β indicating the position in the row. The respective ranges for α and β are (1,25) and (1,40), respectively. In choosing α we may utilize numbers from the range (1,100), taking the remainder modulo 25. Thus, for example, 67 will be considered as 67 - 50 = 17, symbolically 67 $\to$ 17. In choosing β, we may utilize numbers from the range (1,80), taking again remainders modulo 40. Here is one actual completion of the above sampling program:

First draw: $\alpha = 57 \to 07$, $\beta = 36$, $\xi = 6$. The unit in the 7th row and the 36th position has $x_{(7,36)} = 6$, and since $\xi = 6 = x_{(7,36)} = 6$, the unit is accepted.

Second draw: $\alpha = 36 \to 11$, $\beta = 56 \to 16$, $\xi = 8$. The unit with the position (11,16) has $x_{(11,16)} = 4$, and since $\xi = 8 > x_{(11,16)} = 4$, the unit is rejected. We repeat the procedure:

$$\alpha = 65 \to 15 \qquad \beta = 24 \qquad \xi = 2$$

$x_{(15,24)} = 6$; the unit is accepted.

Third draw: $\alpha = 42 \to 17$, $\beta = 61 \to 21$, $\xi = 5$; $x_{(17,21)} = 7$; the unit is accepted.

Thus a rejective sample of size n = 100 is obtained by adding to Table 6.2 three units with positions (7,36), (15,24), and (17, 21) carrying number 6, 6, and 7, respectively.

Example 7.6 Let us correct the Poisson sample of Table 6.4. Since $n' = 102 > 100 = n$, we should remove from s' two units by rejective sampling with probabilities proportionate to $2399 - 100x_i^{.6}$. Since the numbers depend on x_i only very slightly, we can replace this sampling by simple random sampling. Thus we select two random integers in the range (1,102). We have drawn $\xi_1 = 467 \to 067$ and $\xi_2 = 805 \to 005$. Numbering the unit in Table 6.4 in such a way that we run through the first row and then through the second row, etc., until we reach the last row, then the unit indexed by 067 carries digit 4 and the unit indexed by 005 carries digit 5 in Table 6.4. These two digits are underlined in Table 6.4. Removing these two digits we get an approximate rejective sample of size n = 100 with π_i terms proportionate to $x_i^{.6}$.

PROBLEMS AND NOTES

7.1 Let us consider two Poisson samplings from a population of $N = 10$. The first has the probabilities $p_1 = \cdots = p_5 = 1/3$ and $p_6 = \cdots = p_{10} = 1/5$; the second has the probabilities $p'_1 = \cdots = p'_5 = 1/2$ and $p'_6 = \cdots = p'_{10} = x$. Establish x in such a manner that the corresponding conditional sampling given size $s = n$ coincides for both the Poisson samplings.

7.2 Estimate the probability that a sample of size 20 with replacement from the population will not be rejected, if the α_i terms are chosen so that the π_i terms of the corresponding rejective sampling are proportionate to the x_i terms of Table 6.1.

7.3 Perform rejective sampling of size $n = 20$ from the population of Table 6.1 with the π_i terms proportionate to the x_i terms. Realize the sampling as conditional sampling with replacement with drawing probabilities given by (7.31) where $\pi_i = 20x_i/4565$, or more simply, use (7.33). You may make use of Lahiri's method explained in Example 7.5, where single draws with unequal probabilities were realized as draws with equal probabilities succeeded by a random rejection rule.

7.4 *Case* $n = 2$ For this sample size we can give exact expressions for π_i and π_{ij}. Prove that for $n = 2$,

$$\pi_{ij} = \frac{2\alpha_i\alpha_j}{1 - \sum_{k=1}^{N} \alpha_k^2} \qquad 1 \le i \ne j \le N \tag{7.34}$$

and

$$\pi_i = \frac{2\alpha_i(1 - \alpha_i)}{1 - \sum_{k=1}^{N} \alpha_k^2} \qquad 1 \le i \le N \tag{7.35}$$

holds. Further show that

$$\pi_i\pi_j - \pi_{ij} = \frac{2\alpha_i\alpha_j}{\left(1 - \sum_{k=1}^{N} \alpha_k^2\right)^2}\left[(1 - \alpha_i - \alpha_j)^2 + \sum_{k\ne i,j} \alpha_k^2\right] \ge 0$$

7.5 Consider Poisson sampling with probabilities $p_1, \ldots, p_N$. Denote $n = \Sigma_{i=1}^N p_i$, $d = \Sigma_{i=1}^N p_i(1 - p_i)$. Generalize Theorem 7.3 as follows:

$$\pi_i' = P(I_i = 1 | K = k)$$

$$= p_i \left[1 + \frac{(k - n)(1 - p_i)}{d} + \frac{(k - n)^2 - d}{d^2} \times (1 - p_i)(\bar{\bar{p}} - p_i) + o(d^{-1})\right] \qquad 1 \le i \le N \tag{7.36}$$

where k = size s and $o(d^{-1})d \to 0$ as $d \to \infty$ provided that $|k - n|d^{-1/2}$ remains bounded by a constant independent of d.

7.6 Assume that n is fixed, $N \to \infty$, and $N \max_{1 \le i \le N} \pi_i$ remains bounded. Then, under rejective sampling,

$$\alpha_i = p_i\left[1 + (n - 1)p_i - (n - 1) \sum_{j=1}^{N} p_j^2\right] + O(N^{-3}) \tag{7.37}$$

where $p_i = \pi_i/n$, $1 \le i \le N$. Also the remaining forms of formula (7.32) have the same order of precision as (7.37).

Furthermore,

$$\pi_i\pi_j - \pi_{ij} = \frac{\pi_i\{1 - [(n - 1)/n]\pi_i\}\pi_j\{1 - [(n - 1)/n]\pi_j\}}{\sum_{k=1}^N \pi_k\{1 - [(n - 1)/n]\pi_k\}} + O(N^{-4}) \tag{7.38}$$

and the same order of precision may be attached to (7.28), but not to (7.27) (Rao, 1963).

7.7 Under assumptions of Problem 7.6, we have for rejective sampling

$$\pi_i\pi_j - \pi_{ij} = \frac{\pi_i(1 - \lambda_i)\pi_j(1 - \lambda_j)}{d}\left[1 - 2\,\frac{n - 1}{n^2}(\pi_i - \bar{\bar{\pi}})(\pi_j - \bar{\bar{\pi}})\right] + O(N^{-5}) \tag{7.39}$$

where $d = \Sigma_{i=1}^{N} \pi_i(1 - \lambda_i)$, $1 - \lambda_i$ is defined by (3.24) and $\bar{\bar{\pi}} = \Sigma_{i=1}^{N} \pi_i^2(1 - \lambda_i)/d$.

7.8 The best linear predictor of I_i by means of K = size s in Poisson sampling equals

$$\hat{\pi}_i = p_i + \frac{p_i(1 - p_i)}{d}(K - n)$$

where n = EK. The *residual* covariance of I_i and I_j then equals

$$\mathrm{Cov}(I_i - \hat{\pi}_i, I_j - \hat{\pi}_j) = -\frac{p_i(1 - p_i)p_j(1 - p_j)}{d}$$

However this is, asymptotically, the conditional covariance $\mathrm{Cov}(I_i, I_j | K = n)$, i.e., $\pi_{ij} - \pi_i\pi_j$ under the respective rejective sampling. Explain this coincidence!

7.9 Compute quadratic regression of I_i on K and show that it coincides with the right-hand side of (7.36), if we neglect $o(d^{-1})$.

7.10 We have

$$\pi_i\pi_j - \pi_{ij} = \frac{\pi_i(1 - \pi_i)\pi_j(1 - \pi_j)}{d}[1 + o(d^{-1})] \tag{7.40}$$

where $d\, o(d^{-1}) \to 0$ uniformly in $1 \leq i \neq j \leq N$.

Hint: (7.23) implies $d\, o(d^{-1})/(\pi_i\pi_j) \to 0$; considering the events $I_i(1 - I_j) = 1$, $I_j(1 - I_i) = 1$, and $(1 - I_i)(1 - I_j) = 1$, we can similarly derive $d\, o(d^{-1})/[\pi_i(1 - \pi_j)] \to 0$, $d\, o(d^{-1})/[\pi_j(1 - \pi_i)] \to 0$, and $d\, o(d^{-1})/[(1 - \pi_i)(1 - \pi_j)] \to 0$.

8

SAMPFORD-DURBIN MODIFICATION OF REJECTIVE SAMPLING

We have seen that in rejective sampling the drawing probabilities (7.31) provide desired probabilities of inclusion π_i approximately to a satisfactory degree of precision. If $d = \Sigma_{i=1}^{N} \pi_i(1 - \pi_i)$ is not very small, we may even use a simpler formula, namely

$$\alpha_i = \frac{\pi_i}{1 - \pi_i} \qquad 1 \leq i \leq N \tag{8.1}$$

However, there is a slight modification of rejective sampling that yields the inclusion probabilities π_i exactly; it was suggested by Durbin (1967) for n = 2 and by Sampford (1967) in the general case.

SAMPFORD-DURBIN MODIFICATION OF REJECTIVE SAMPLING We select the first unit with drawing probabilities

$$\alpha_i^{I} = \frac{1}{n} \pi_i \qquad 1 \leq i \leq N \tag{8.2}$$

The remaining n - 1 units are drawn with drawing probabilities

$$\alpha_i^{II} = \lambda \frac{\pi_i}{1 - \pi_i} \qquad 1 \leq i \leq N . \tag{8.3}$$

If the units drawn are all distinct, we accept the sample; otherwise we reject it as soon as the first duplicate appears and start again.

THEOREM 8.1 In Sampford-Durbin sampling the inclusion probabilities equal exactly π_i if (8.2) and (8.3) hold.

Proof. Since $\Sigma_{i=1}^{N} \pi_i = n$, we have

$$\sum_{i \in s} (1 - \pi_i) = \sum_{i \in r} \pi_i \qquad \text{if size } s = n \tag{8.4}$$

where $r = S - s$. Now the probability of a sample s, provided that the element k was drawn as the first, equals

$$P(s, \text{the first element was } k) = c\pi_k \prod_{\substack{j \in s \\ j \neq k}} \frac{\pi_j}{1 - \pi_j}$$

$$= c(1 - \pi_k) \prod_{j \in s} \frac{\pi_j}{1 - \pi_j}$$

Altogether

$$P(s) = c \sum_{k \in s} (1 - \pi_k) \prod_{j \in s} \frac{\pi_j}{1 - \pi_j} = c \sum_{k \in r} \pi_k \prod_{j \in s} \lambda_j$$

where we have utilized (8.4) and denoted $\lambda_j = \pi_j/(1 - \pi_j)$. Take a sample s including i and denote by s_{ik} the sample obtained from s by replacing i by some $k \in r$. Then

$$P(s) = c\lambda_i \sum_{k \in r} (1 - \pi_k) \prod_{j \in s_{ik}} \lambda_j$$

Now, denoting by $\Sigma_{r \ni i}$ the summation over all samples s that do not include i, we obtain, after consideration,

$$\sum_{s \ni i} P(s) = \lambda_i c \sum_{s \ni i} \sum_{k \in r} (1 - \pi_k) \prod_{j \in s_{ik}} \lambda_j$$

$$= \lambda_i c \sum_{r \ni i} \sum_{h \in s} (1 - \pi_h) \prod_{j \in s} \lambda_j$$

$$= \lambda_i \sum_{r \ni i} P(s)$$

$$= \frac{\pi_i}{1 - \pi_i} \left[1 - \sum_{s \ni i} P(s)\right]$$

Consequently, $\Sigma_{s \ni i} P(s) = \pi_i$. Q.E.D.

Remark 8.1 Comparing (7.31) with (8.1) and (8.2), we can see that the probabilities α_i are a weighted average of the probabilities α_i^{I} and α_i^{II}; the weights, however, are not 1 and n - 1 but 1 and $d - 1 + \bar{\bar{\pi}}$. The difference between rejective sampling based on (7.31) and the Sampford-Durbin sampling based on (8.1) and (8.2) will be for practical purposes negligible.

Theorem 7.4 is true also for the Sampford-Durbin sampling:

THEOREM 8.2 For the Sampford-Durbin modification of rejective sampling let π_i and π_{ij} be the inclusion probabilities. Then

$$\pi_{ij} = \pi_i \pi_j \left[1 - \frac{(1 - \pi_i)(1 - \pi_j)}{d} + o(d^{-1})\right] \tag{8.5}$$

where $d = \Sigma_{i=1}^{N} \pi_i(1 - \pi_i)$, and $d\, o(d^{-1}) \to 0$ uniformly in $1 \le i \le N$.

Proof. Introduce

$$\lambda_j = \frac{\pi_j}{1 - \pi_j} \qquad 1 \le j \le N \tag{8.6}$$

and consider Sampford-Durbin sampling with probabilities

$$Q(s) = c\left\{\sum_{k \in r} \pi_k\right\} \prod_{j \in s} \lambda_j \qquad s \subset S \qquad \text{size } s = n \tag{8.7}$$

and rejective sampling with probabilities

$$P(s) = c^* \prod_{i \in s} \lambda_i \qquad s \subset S \qquad \text{size } s = n \tag{8.8}$$

The expectations via P and Q will be denoted $E^P(\cdot)$, $E^Q(\cdot)$, respectively. Obviously, the likelihood ratio for Q via P equals

$$\frac{Q(s)}{P(s)} = c_1 \sum_{k \in r} \pi_k = c_1 \sum_{i=1}^{N} \pi_i(1 - I_i) = \frac{\sum_{i=1}^{N} \pi_i(1 - I_i)}{\sum_{i=1}^{N} \pi_i(1 - \varkappa_i)}$$

where I_i are the inclusion indicators and

$$\varkappa_i = E^P(I_i) \qquad 1 \le i \le N \tag{8.9}$$

are the *exact* inclusion probabilities corresponding to the rejective sampling $P(\cdot)$. Theorem 7.3 entails

$$\pi_i = \varkappa_i\left[1 + \frac{(\bar{\bar{\varkappa}} - \varkappa_i)(1 - \varkappa_i)}{d_*} + o(d_*^{-1})\right] \tag{8.10}$$

where

$$\bar{\bar{\varkappa}} = \frac{\Sigma_{i=1}^N \varkappa_i^2(1 - \varkappa_i)}{\Sigma_{i=1}^N \varkappa_i(1 - \varkappa_i)} \qquad d_* = \sum_{i=1}^N \varkappa_i(1 - \varkappa_i) \tag{8.11}$$

[The role of (π_i, p_i) is played by $(\varkappa_i, \pi_i)$ in our case.] Thus, under Q,

$$\begin{aligned}\pi_{ij} = E^Q(I_iI_j) &= \frac{E^P[\Sigma_{k=1}^N \pi_k(1 - I_k)I_iI_j]}{\Sigma_{k=1}^N \pi_k(1 - \varkappa_k)} \\ &= \frac{\Sigma_{k\neq i,j} \pi_k(\varkappa_{ij} - \varkappa_{ijk})}{\Sigma_{k=1}^N \pi_k(1 - \varkappa_k)}\end{aligned} \tag{8.12}$$

where

$$\varkappa_{ij} = E^P(I_iI_j) \qquad \varkappa_{ijk} = E^P(I_iI_jI_k)$$

Probabilities $\varkappa_{ijk}/\varkappa_i$ and $\varkappa_{ik}/\varkappa_i$ may be viewed as second- and first-order inclusion probabilities for rejective sampling condition on $I_i = 1$. Since rejective sampling conditional on $I_i = 1$, i.e., $s \ni i$, is again a rejective sampling, Theorem 7.4 is applicable. Thus

$$\frac{\varkappa_{ijk}}{\varkappa_i} = \frac{\varkappa_{ij}\varkappa_{ik}}{\varkappa_i^2}\left[1 - \frac{(1 - \varkappa_{ij}/\varkappa_i)(1 - \varkappa_{ik}/\varkappa_i)}{d_1} + o(d_1^{-1})\right] \tag{8.13}$$

where

$$d_1 = \sum_{j\neq i} \frac{\varkappa_{ij}}{\varkappa_i}\left(1 - \frac{\varkappa_{ij}}{\varkappa_i}\right)$$

Theorem 7.4, however, may also be applied to probabilities $\varkappa_{ij}$, yielding

$$\varkappa_{ij} = \varkappa_i\varkappa_j\left[1 - \frac{(1-\varkappa_i)(1-\varkappa_j)}{d_*} + o(d_*^{-1})\right] \qquad 1 \le i \ne j \le N \tag{8.14}$$

where

$$d_* = \sum_{i=1}^{N} \varkappa_i(1-\varkappa_i) \tag{8.15}$$

Thus $d_1 = d_* + O(1)$ and

$$\varkappa_{ijk} = \varkappa_i\varkappa_j\varkappa_k \times \left[1 - \frac{(1-\varkappa_i)(1-\varkappa_j) + (1-\varkappa_i)(1-\varkappa_k) + (1-\varkappa_j)(1-\varkappa_k)}{d} + o(d_*^{-1})\right] \tag{8.16}$$

Consequently

$$\begin{aligned}
&\sum_{k\ne i,j}^{N} \pi_k(\varkappa_{ij} - \varkappa_{ijk}) \\
&= \sum_{k\ne i,j}^{N} \pi_k\Bigg\{\varkappa_i\varkappa_j - \frac{\varkappa_i\varkappa_j(1-\varkappa_i)(1-\varkappa_j)}{d_*} - \varkappa_i\varkappa_j\varkappa_k \\
&\quad + \varkappa_i\varkappa_j\varkappa_k\left[\frac{(1-\varkappa_i)(1-\varkappa_j) + (1-\varkappa_i)(1-\varkappa_k) + (1-\varkappa_j)(1-\varkappa_k)}{d_*}\right] \\
&\quad + o(d_*^{-1})[\varkappa_i\varkappa_j(1-\varkappa_k)]\Bigg\} = \varkappa_i\varkappa_j \sum_{k\ne i,j} \pi_k(1-\varkappa_k) \\
&\quad - \varkappa_i\varkappa_j(1-\varkappa_i)(1-\varkappa_j)d_*^{-1} \sum_{k\ne i,j} \pi_k(1-\pi_k) \\
&\quad + \varkappa_i\varkappa_j(1-\varkappa_i + 1 - \varkappa_j)d_*^{-1} \sum_{k\ne i,j}^{N} \varkappa_k\pi_k(1-\varkappa_k) \\
&\quad + o(d_*^{-1}) \sum_{k\ne i,j}^{N} \pi_k(1-\varkappa_k)
\end{aligned}$$

$$= \varkappa_i \varkappa_j \sum_{k=1}^{N} \pi_k (1 - \varkappa_k)[1 - (1 - \varkappa_i)\varkappa_i d_*^{-1} - (1 - \varkappa_j)\varkappa_j d_*^{-1}$$

$$- (1 - \varkappa_i)(1 - \varkappa_j) d_*^{-1} + (1 - \varkappa_i + 1 - \varkappa_j) d_*^{-1} \bar{\bar{\varkappa}} + o(d_*^{-1})]$$

$$= \left[\sum_{k=1}^{N} \pi_k (1 - \varkappa_k)\right] \Big\{ \varkappa_i \varkappa_j [1 + (\bar{\bar{\varkappa}} - \varkappa_i)(1 - \varkappa_i) d_*^{-1}]$$

$$\times [1 + (\bar{\bar{\varkappa}} - \varkappa_j)(1 - \varkappa_j) d_*^{-1}] - \varkappa_i \varkappa_j [(1 - \varkappa_i)(1 - \varkappa_j) d_*^{-1}$$

$$+ o(d_*^{-1})] \Big\}$$

$$= \left[\sum_{k=1}^{N} \pi_k (1 - \varkappa_k)\right] \pi_i \pi_j [1 - (1 - \pi_i)(1 - \pi_j) d^{-1} + o(d^{-1})] \tag{8.17}$$

We have utilized the fact that $d_* = d + O(1)$ and $\pi_i = \varkappa_i + O(d^{-1})$.

Now (8.12) and (8.17) yield

$$\pi_{ij} = \frac{\Sigma_{k \neq i,j}^{N} \pi_k (\varkappa_{ij} - \varkappa_{ijk})}{\Sigma_{k=1}^{N} \pi_k (1 - \varkappa_k)}$$

$$= \pi_i \pi_j [1 - (1 - \pi_i)(1 - \pi_j) d^{-1} + o(d^{-1})] \qquad \text{Q.E.D.}$$

TABLE 8.1 Exact values of $\pi_i \pi_j - \pi_{ij}$ for a Sampford-Durbin sampling

i \ j	2	3	4	5	6	7	8	9	10
1	.0088	.0098	.0110	.0113	.0113	.0111	.0106	.0087	.0074
2		.0247	.0277	.0283	.0283	.0274	.0259	.0211	.0178
3			.0304	.0310	.0310	.0299	.0283	.0229	.0194
4				.0343	.0343	.0329	.0309	.0249	.0210
5					.0348	.0332	.0312	.0250	.0210
6						.0332	.0312	.0250	.0210
7							.0294	.0234	.0196
8								.0217	.0183
9									.0145

TABLE 8.2 Values $c_i c_j$ corresponding to the entries in Table 8.1

i \ j	2	3	4	5	6	7	8	9	10
1	.0094	.0102	.0113	.0114	.0114	.0109	.0102	.0082	.0069
2		.0255	.0281	.0284	.0284	.0271	.0255	.0205	.0172
3			.0307	.0311	.0311	.0297	.0279	.0225	.0189
4				.0342	.0342	.0327	.0307	.0247	.0208
5					.0347	.0331	.0311	.0250	.0210
6						.0331	.0311	.0250	.0210
7							.0297	.0239	.0201
8								.0225	.0189
9									.0152

Example 8.1 We shall consider an example due to Sampford (1967). In this example N = 10, n = 5, and $\pi_1 = .90$, $\pi_2 = .70$, $\pi_3 = .65$, $\pi_4 = .55$, $\pi_5 = .50$, $\pi_6 = .50$, $\pi_7 = .40$, $\pi_8 = .35$, $\pi_9 = .25$, $\pi_{10} = .20$. Table 8.1 provides exact (rounded off) values for $\pi_i\pi_j - \pi_{ij}$. Theorem 8.2 suggests that $\pi_i\pi_j - \pi_{ij} \sim c_i c_j$ approximately, and Table 8.2 provides values $c_i c_j$ based on (3.16) and (3.17) or, equivalently, on (3.23) and (3.24).

We can observe a very good agreement between Tables 8.1 and 8.2, good enough for all practical purposes. The maximum absolute difference equals .0008 (i = 2, j = 3), and the maximum relative difference equals 7% (i = 1, j = 2). A table very similar to Table 8.2 would be obtained by using the λ_i terms from (7.28a) in Eq. (7.28).

PROBLEMS AND NOTES

8.1 Carry out a Sampford-Durbin sample of size n = 5 from the population of Table 6.1 with $\pi_i = 5x_i/4565$.

8.2 *Case n = 2* If n = 2, then in Sampford-Durbin sampling

$$\pi_{ij} = \pi_i\pi_j\left(\frac{1}{1-\pi_i} + \frac{1}{1-\pi_j}\right)\left(2 + \sum_{k=1}^{N} \frac{k}{1-\pi_k}\right)^{-1} \qquad 1 \le i \ne j \le N \qquad (8.18)$$

holds.

8.3 Formula (7.13) is applicable to Sampford-Durbin sampling for large n, with $p_i = \pi_i$, $1 \le i \le N$.

8.4 Show that $\pi_i \pi_j - \pi_{ij} > 0$, $1 \le i \ne j \le N$, for Sampford-Durbin sampling of any size (Sampford, 1967).

8.5 Assume that n is fixed, $N \to \infty$, and $N \max_{1 \le i \le N} \pi_i$ remains bounded. Then, under Sampford-Durbin sampling

$$\pi_i \pi_j - \pi_{ij} = \frac{\pi_i(1 - \lambda_i)\pi_j(1 - \lambda_j)}{d}\left[1 - \frac{n - 1}{n^2}(\pi_i - \bar{\bar{\pi}})(\pi_j - \bar{\bar{\pi}})\right] + O(N^{-5}) \tag{8.19}$$

where $d = \Sigma_{i=1}^{N} \pi_i(1 - \lambda_i)$, $1 - \lambda_i$ is given by (3.24), and $\bar{\bar{\pi}} = \Sigma_{i=1}^{N} \pi_i^2(1 - \lambda_i)/d$.

9

SUCCESSIVE SAMPLING

Successive sampling consists of a sequence of independent draws of one unit with some constant drawing probabilities $\alpha_1, \ldots, \alpha_N$, $\Sigma_{i=1}^N \alpha_i = 1$. This sequence of draws is continued until there are n *distinct* units in the sample s. If a draw provides a unit already selected, we ignore *this unit*.

Thus in contrast to rejective sampling, in case of a duplicate, we reject only this duplicate and not the whole partially built sample. The advantage of successive sampling is that the average number of draws will only moderately exceed the sample size. On the other hand, the theory of successive sampling is more complicated than the theory of rejective sampling. A basic contribution in this respect is due to Rosén (1972). In this chapter we shall use an approach which is different from though influenced by Rosén.

THEOREM 9.1 In successive sampling of size n with drawing probabilities $\alpha_1, \ldots, \alpha_N$

$$P(s) = \left(\prod_{i \in s} \alpha_i\right) \sum_{(r_1, \ldots, r_n)}^{s} [(1 - \alpha_{r_1}) \times (1 - \alpha_{r_1} - \alpha_{r_2}) \cdots (1 - \alpha_{r_1} - \cdots - \alpha_{r_{n-1}})]^{-1} \tag{9.1}$$

holds, where $\Sigma^s_{(r_1, \ldots, r_n)}$ denotes the summation over all permutations of members of s.

Proof. The probability of drawing a unit i in the first draw is α_i; the probability of obtaining a unit $j \neq i$ in the nearest next draw when i is not selected is $\alpha_j/(1 - \alpha_i)$; the probability of drawing a unit $k \neq i, j$ in the nearest next draw, when neither i nor j is selected, equals $\alpha_k/(1 - \alpha_i - \alpha_j)$, etc. Thus we obtain a sequence of n distinct units $r_1, \ldots, r_n$ with probability

$$\frac{\alpha_{r_1}\alpha_{r_2}\cdots\alpha_{r_n}}{(1 - \alpha_{r_1})(1 - \alpha_{r_1} - \alpha_{r_2})\cdots(1 - \alpha_{r_1} - \cdots - \alpha_{r_{n-1}})} \tag{9.2}$$

Denoting the *set* consisting of $r_1, \ldots, r_n$ by s, we can also write

$$\alpha_{r_1}\alpha_{r_2}\cdots\alpha_{r_n} = \prod_{i \in s} \alpha_i$$

Now (9.2) is the probability of obtaining s in a certain order. Adding for all n! possible orderings of the units of s, we obtain (9.1). Q.E.D.

Now let T denote the number of draws necessary to obtain n distinct units in the sample.

Further let T_i denote the number of draws necessary to obtain n distinct units in the sample, all of them distinct from i; the draws when i is drawn are not counted.

Finally let T_{ij} denote the number of draws necessary to obtain n distinct units in the sample, all of them distinct from both i and j; the draws when i or j is drawn are not counted.

THEOREM 9.2 Let π_i and π_{ij}, $1 \leq i \neq j \leq N$ be the first- and second-order inclusion probabilities for successive sampling with drawing probabilities $\alpha_1, \ldots, \alpha_N$.

Then

$$1 - \pi_i = E(1 - \alpha_i)^{T_i} \qquad 1 \leq i \leq N \tag{9.3}$$

$$1 - \pi_i - \pi_j + \pi_{ij} = E(1 - \alpha_i - \alpha_j)^{T_{ij}} \qquad 1 \leq i \neq j \leq N \tag{9.4}$$

Proof. If we realize the successive sampling as a sequence of draws which continues until there are n distinct units drawn, we may write

$$1 - \pi_i = \sum_{t=n}^{\infty} \sum_{(i)}^{(t)} \alpha_{J_1} \alpha_{J_2} \cdots \alpha_{J_t}$$

where $\Sigma_{(i)}^{(t)}$ denotes summation over all sequences $(J_1, \ldots, J_t)$ of units from S such that

1. The length of the sequence is t.
2. The sequence does not include i.
3. There are n distinct units among $J_1, \ldots, J_t$ and J_t is distinct from the rest.

Now, we may equivalently write

$$1 - \pi_i = \sum_{t=n}^{\infty} (1 - \alpha_i)^t \sum_{(i)}^{(t)} \frac{\alpha_{J_1}}{(1 - \alpha_i)} \frac{\alpha_{J_2}}{(1 - \alpha_i)} \frac{\alpha_{J_t}}{(1 - \alpha_i)}$$

However, we can easily see that

$$\sum_{(i)}^{(t)} \frac{\alpha_{J_1}}{1 - \alpha_i} \cdots \frac{\alpha_{J_t}}{1 - \alpha_i} = P(T_i = t)$$

Thus

$$1 - \pi_i = \sum_{t=n}^{\infty} (1 - \alpha_i)^t P(T_i = t) = E(1 - \alpha_i)^{T_i}$$

Equation (9.4) may be proved in a similar way, if we realize that $1 - \pi_i - \pi_j + \pi_{ij}$ denotes the probability that neither i nor j is in the sample.

If we replace T_i by a common constant t, we get approximately that

$$1 - \pi_i \doteq (1 - \alpha_i)^t \qquad 1 \le i \le N \tag{9.5}$$

Since $\Sigma_{i=1}^{N}(1 - \pi_i) = N - n$, t should satisfy

$$N - n = \sum_{i=1}^{N} (1 - \alpha_i)^t \tag{9.6}$$

Q.E.D.

Example 9.1 If $N = 10$, $n = 4$, $\alpha_i = i/55$, $1 \le i \le 10$, then the solution of (9.6) equals

$$t = 5.191$$

It was obtained by iterating the recurrent relation

$$t_{n+1} = \sum_{i=1}^{10} (1 - \alpha_i)^{t_n} + t_n - 6$$

A comparison of exact values of the π_i terms with those obtained from (9.5) is given in Table 9.1. We observe a fairly good agreement, the maximum relative error being 1%. To the same order of approximation we could also put

$$1 - \pi_i = e^{-\alpha_i t} \qquad 1 \le i \le N \tag{9.7}$$

where t satisfies

$$N - n = \sum_{i=1}^{N} e^{-\alpha_i t} \tag{9.8}$$

TABLE 9.1 Exact π_i values and approximations obtained from (9.5) for successive sampling ($N = 10$, $n = 4$, $\alpha_i = i/55$)

i	π_i	(9.5)	Difference	Relative difference (%)
1	.0902	.0908	.0006	.7
2	.1747	.1749	.0002	.1
3	.2534	.2526	-.0008	.3
4	.3262	.3243	-.0019	.6
5	.3930	.3903	-.0027	.7
6	.4537	.4510	-.0027	.6
7	.5086	.5067	-.0019	.4
8	.5578	.5578	.0000	.0
9	.6017	.6045	.0028	.5
10	.6408	.6471	.0063	1.0

From (9.5) it follows that

$$\alpha_i \doteq 1 - (1 - \pi_i)^{1/t} \doteq \frac{1}{t}\pi_i + \frac{1}{2}\frac{t-1}{t^2}\pi_i^2$$

$$\doteq \frac{1}{t}\frac{\pi_i}{1 - (1/2)[(t-1)/t]\pi_i} \doteq \frac{\pi_i}{n}\left[1 + \frac{1}{2}\frac{t-1}{t}(\pi_i - \bar{\pi})\right]$$

$$\doteq \frac{\pi_i}{n}\left[1 + \frac{1}{2}\frac{n-1}{n}(\pi_i - \bar{\pi})\right] \tag{9.9}$$

where $\bar{\pi} = \Sigma_{i=1}^{N} \pi_i^2/n$.

Of course, in applications it often suffices to put $\alpha_i = \lambda\pi_i(1 + (1/2)\pi_i)$, or $\alpha_i = \lambda\pi_i(1 - (1/2)\pi_i)^{-1}$, or even $\alpha_i = \pi_i/n$.

If we compare (9.9) with (7.32) for rejective sampling we can see that the approximate α_i terms are more proportionate to the π_i terms than in rejective sampling. Also the range of variability of the π_i terms is larger for successive sampling than for rejective sampling if the same α_i terms are used. Actually,

$$\frac{\max \alpha_i}{\min \alpha_i} \geq \frac{\max \pi_i^{(s)}}{\min \pi_i^{(s)}} \geq \frac{\max \pi_i^{(r)}}{\min \pi_i^{(r)}} \tag{9.10}$$

where $\pi_i^{(s)}$ and $\pi_i^{(r)}$ refer to the successive and rejective sampling with $\alpha_1, \ldots, \alpha_N$, respectively. The cases of equality occur only if $\alpha_1 = \cdots = \alpha_N$. Inequalities (9.10) are based on approximate formulas for the π_i terms and on numerical experience. We do not have an exact proof.

Example 9.2 If N = 10, n = 4, and $\alpha_i = i/55$, then

$$\frac{\max \alpha_i}{\min \alpha_i} = \frac{10}{1} = 10$$

$$\frac{\max \pi_i^{(s)}}{\min \pi_i^{(s)}} = \frac{.6408}{.0902} = 7.1$$

$$\frac{\max \pi_i^{(r)}}{\min \pi_i^{(r)}} = \frac{.0599}{.01070} = 5.6$$

Formula (9.5) evaluates the π_i terms with a remainder of order 1/N. This is not sufficient for finding an asymptotic expression for $\pi_i\pi_j - \pi_{ij}$, because these covariances themselves are of order 1/N. Thus we must try to evaluate π_i as well as π_{ij} with a remainder of order $o(1/N)$. This will be done under the following conditions:

Assume that

$$N \to \infty \tag{9.11}$$

and that n grows also in such a way that

$$\frac{n}{N} \to \lambda \qquad 0 < \lambda < 1 \tag{9.12}$$

Furthermore, there is a positive $\varepsilon > 0$, independent of N, such that

$$\varepsilon \le \frac{\min \alpha_i}{\max \alpha_i} \tag{9.13}$$

Now remember that T denotes the number of draws necessary to obtain n distinct units.

THEOREM 9.3 Under conditions (9.11) through (9.13), we have

$$ET = t + o(\sqrt{n}) \tag{9.14}$$

$$\operatorname{Var} T = \left[\frac{\sum_{i=1}^{N} e^{-\alpha_i t}(1 - e^{-\alpha_i t})}{\left(\sum_{i=1}^{N} \alpha_i e^{-\alpha_i t}\right)^2} - t\right][1 + o(1)] \tag{9.15}$$

where t is defined by (9.8).

Proof. See Rosén (1970).

Now let us sketch our plan for evaluating $\pi_i\pi_j - \pi_{ij}$. First denote

$$ET = t_0 \qquad ET_i = t_i \qquad ET_{ij} = t_{ij} \qquad 1 \le i \ne j \le N \tag{9.16}$$

and

$$\Delta_i = t_i - t_0 \qquad \text{and} \qquad \Delta_{ij} = t_{ij} - t_0$$

$$1 \le i \ne j \le N \tag{9.17}$$

Then making use of the Taylor development

$$(1 - \alpha_i)^{T_i} = (1 - \alpha_i)^{t_i}\Big[1 - \alpha_i(T_i - t_i) + \frac{1}{2}(T_i - t_i)(T_i - t_i - 1)\alpha_i^2 + \text{remainder}\Big] \tag{9.18}$$

we obtain from (9.3)

$$\begin{aligned} 1 - \pi_i &= (1 - \alpha_i)^{t_i}\left[1 + \frac{1}{2}\alpha_i^2 \text{ var } T_i + o\left(\frac{1}{N}\right)\right] \\ &= \exp\left[-\alpha_i t_i + \frac{1}{2}\alpha_i^2 \text{ var } T - \frac{1}{2}\alpha_i^2 t_i + o\left(\frac{1}{N}\right)\right] \\ &= \exp\left[-\alpha_i t_0 - \Delta_i \alpha_i + \frac{1}{2}\alpha_i^2(\text{var } T - t_0) + o\left(\frac{1}{N}\right)\right] \end{aligned} \tag{9.19}$$

provided that the replacement of var T_i by var T and of $\alpha_i^2 t_i$ by $\alpha_i^2 t_0$ is justified.

Similarly, we evaluate from (9.4)

$$1 - \pi_i - \pi_j + \pi_{ij} = \exp\Big[-(\alpha_i + \alpha_j)t_0 - \Delta_{ij}(\alpha_i + \alpha_j) + \frac{1}{2}(\alpha_i + \alpha_j)^2(\text{var } T - t_0) + o\left(\frac{1}{N}\right)\Big] \tag{9.20}$$

Consequently

$$\begin{aligned} \pi_i\pi_j - \pi_{ij} &= (1 - \pi_i)(1 - \pi_j) - (1 - \pi_i - \pi_j + \pi_{ij}) \\ &= (1 - \pi_i)(1 - \pi_j)\Big\{1 - \exp\Big[-\Delta_{ij}(\alpha_i + \alpha_j) + \Delta_j\alpha_j + \alpha_i\alpha_j(\text{var } T - t_0) + o\left(\frac{1}{N}\right)\Big]\Big\} \\ &= (1 - \pi_i)(1 - \pi_j)\Big[\Delta_{ij}(\alpha_i + \alpha_j) - \Delta_i\alpha_i + \Delta_j\alpha_j - \alpha_i\alpha_j(\text{var } T - t_0) + o\left(\frac{1}{N}\right)\Big] \end{aligned} \tag{9.21}$$

If we, in addition, show that

$$\Delta_{ij} = \Delta_i + \Delta_j + o(1) \tag{9.22}$$

we can also write

$$\pi_i\pi_j - \pi_{ij} = (1 - \pi_i)(1 - \pi_j)\left[\Delta_i\alpha_j + \Delta_j\alpha_i - \alpha_i\alpha_j(\text{var } T - t_0) + o\left(\frac{1}{N}\right)\right] \tag{9.23}$$

In order to make this procedure rigorous, we have to solve a few technical problems. First we shall study the Δ_i terms. Consider the space of infinite sequences $(J_1, J_2, \ldots)$ of draws and for each sequence define T and T_i. Obviously, we may write

$$T_i = T - \xi_i + \eta_i \tag{9.24}$$

where ξ_i denotes the number of draws with ordinal number $k \leq T$ and such that $J_k = i$, and

$$\eta_i = 0 \qquad \text{if } \xi_i = 0$$

and

η_i = the number of draws following the Tth draw necessary to select a unit distinct from all the units $J_1, \ldots, J_T$ if $\xi_i > 0$; the draws when i is drawn are not counted.

Thus $\Delta_i = E\eta_i - E\xi_i$. If the units $J_1, \ldots, J_T$ are fixed and such that $\xi_i > 0$, the conditional distribution of η_i is geometric,

$$P(\eta_i = k | J_1, \ldots, J_T) = A^{k-1}(1 - A)$$

where

$$A = \frac{\Sigma_{j \in s}(\alpha_j - \alpha_i)}{1 - \alpha_i} \tag{9.25}$$

and s denotes the sample consisting of n distinct units among $J_1, \ldots, J_T$. Thus

$$E(\eta_i | J_1, \ldots, J_T) = \begin{cases} \dfrac{1}{1 - A} & \text{if } \xi_i > 0 \\ 0 & \text{if } \xi_i = 0 \end{cases} \tag{9.26}$$

Assumption (9.13) entails that $\alpha_i \geq \varepsilon/N$, so that, for N sufficiently large, in view of (9.12),

$$\frac{1}{1 - A} \le \frac{N}{(N - n)\varepsilon} \le \frac{2}{(1 - \lambda)\varepsilon}$$

This yields

$$0 \le E\eta_i \le \frac{2}{(1 - \lambda)\varepsilon} \qquad 1 \le i \le N < \infty \tag{9.27}$$

As for ξ_i, we can write

$$\xi_i = \sum_{k=1}^{\infty} \xi_{ik}$$

where

$$\xi_{ik} = \begin{cases} 1 & \text{if } J_k = i \qquad T \ge k \\ 0 & \text{otherwise} \end{cases}$$

Thus

$$E\xi_i = \sum_{k=1}^{\infty} E\xi_{ik} = \sum_{k=1}^{\infty} P(J_k = i,\ T \ge k)$$

The events $\{T < k\}$ and $\{J_k = i\}$, however, are independent, so that $\{T \ge k\}$ and $\{J_k = i\}$ are also independent, and

$$E\xi_i = P(J_k = i) \sum_{k=1}^{\infty} P(T \ge k) = \alpha_i ET \tag{9.28}$$

Now (9.13) and (9.8) entail that for large N,

$$\frac{1}{2}(1 - \lambda) < e^{-\varepsilon t/N}$$

so that t is of the same order as N. Since ET is of the same order as t, in view of (9.14), and

$$\alpha_i \le \frac{1}{\varepsilon}\frac{1}{N} \tag{9.29}$$

in view of (9.13), we can assert that the numbers $E\xi_i$ are uniformly bounded as $N \to \infty$. Thus we have obtained the following:

LEMMA 9.4 Under assumptions (9.11) through (9.13) the numbers Δ_i of (9.17) are uniformly bounded.

Now it should be observed that the sampling setup in which the unit i is deleted and the α_j terms are replaced by $\alpha_j/(1 - \alpha_i)$, $j \neq i$, satisfies also conditions (9.11) through (9.13), possibly with a different $\varepsilon > 0$. However in this new setup the relation between T_{ij} and T_i is the same as between T_j and T in the original setup. Thus Lemma 9.4 applies to $ET_{ij} - ET_i = \Delta_{ij} - \Delta_i$ as well, and we have the following:

LEMMA 9.5 Under conditions (9.11) through (9.13) the numbers Δ_{ij} of (9.17) are uniformly bounded.

Lemma 9.5 justifies the replacement of $\alpha_i^2 t_i$ by $\alpha_i^2 t_0$ in (9.19), and of $(\alpha_i + \alpha_j)^2 t_{ij}$ by $(\alpha_i + \alpha_j)^2 t_0$ in (9.20).

Next we need to show that Var T_i and Var T_{ij} can be replaced by Var T. If we denote by t_i' the solution of (9.8) for the case when the unit i is removed and the α_j terms are replaced by $\alpha_j/(1 - \alpha_i)$, that is, for

$$N - n - 1 = \sum_{j \neq i}^{N} \exp\left(- \frac{\alpha_j t_i'}{1 - \alpha_i}\right) \tag{9.30}$$

we can easily see that

$$\frac{t_i'}{1 - \max_{1 \leq j \leq N} \alpha_j} \leq t \leq t_i'\left(1 + 2 \max_{i \leq j \leq N} \alpha_j\right) \tag{9.31}$$

where t is the solution of (9.8). However, (9.31) and (9.15) imply that the following lemma is true:

LEMMA 9.6 Under assumptions (9.11) through (9.13),

$$\text{Var } T_i = [1 + o(1)]\text{Var } T \tag{9.32}$$

and

$$\text{Var } T_{ij} = [1 + o(1)]\text{Var } T \tag{9.33}$$

hold.

The part of the lemma concerning T_{ij} can be proved by the fact that, again, the relation between T_{ij} and T_i is analogous to one between T_j and T.

COROLLARY 9.7 Under assumptions (9.11) through (9.13) the relations (9.18) through (9.21) are justified, and there is a constant K independent of N such that

$$|\pi_i \pi_j - \pi_{ij}| \leq \frac{K}{N} \qquad 1 \leq i \neq j \leq N \tag{9.34}$$

Proof. The proof follows from the above lemmas.

Now we are in a position to analyze the numbers Δ_i and Δ_{ij} more explicitly. Equation (9.24) entails

$$\Delta_i = E\eta_i - E\xi_i \tag{9.35}$$

Next $E\xi_i$ was established in (9.28), and (9.25) and (9.26) imply

$$E\eta_i = \pi_i E\left(\frac{1}{1 - A}\middle| s \ni i\right) \tag{9.36}$$

where A is as in (9.25), and $E(\cdot | s \ni i)$ denotes the conditional expectation over samples including the unit i. Observe that the events $\{s \ni i\}$ and $\{s \ni j\}$ are asymptotically independent in view of (9.34), i.e.,

$$\left(\frac{\pi_{ij}}{\pi_i} - \pi_j\right) \to 0 \tag{9.37}$$

Consequently, in view of (9.25),

$$\begin{aligned} E(1 - A | s \ni i) &= [1 + o(1)]E(1 - A) \\ &= [1 + o(1)]E\left[\sum_{i=1}^{N} \alpha_i (1 - I_i)\right] = [1 + o(1)] \sum_{i=1}^{N} \alpha_i (1 - \pi_i) \\ &= \sum_{i=1}^{N} \alpha_i (1 - \pi_i) + o(1) \end{aligned}$$

Furthermore, denoting $E(1 - A | s \ni i) = a$, we have

$$E\left(\frac{1}{1 - A}\middle| s \ni i\right) = \frac{1}{a} + E\left[\frac{(A - a)^2}{a^3} \xi \middle| s \ni i\right]$$

where ξ is a bounded random variable. Now since $a \geq \delta > 0$ for some δ independent of N, in order to show that

$$E\left(\frac{1}{1 - A}\middle| s \ni i\right) = \frac{1}{a} + o(1) \tag{9.38}$$

it remains to show that

$$E[(A - a)^2 | s \ni i] \to 0 \tag{9.39}$$

Since the conditional on $s \ni i$ setup satisfies the same basic conditions (9.11) through (9.13), it is sufficient to show that

$$E[(A - a)^2] \to 0 \tag{9.40}$$

However, as we shall see later in Chapter 17,

$$E(A - a)^2 = \frac{1}{2} \sum_{i=1}^{N} \sum_{j=1}^{N} (\alpha_i - \alpha_j)^2 (\pi_i \pi_j - \pi_{ij})$$

Consequently, in view of (9.34) and (9.29) the right side converges to 0. Combining (9.28), (9.35), (9.36), and (9.38), we obtain the following:

LEMMA 9.8 Under assumptions (9.11) through (9.13), we have

$$\Delta_i = \frac{\pi_i}{\sum_{j=1}^{N} \alpha_j (1 - \pi_j)} - \alpha_i t + o(1) \qquad 1 \leq i \leq N \tag{9.41}$$

where t is the solution of (9.8).

Furthermore, (9.22) is true.

The second statement of the last lemma follows from the representation $\Delta_{ij} = (ET_{ij} - ET_i) + \Delta_i$ and from noting that T_{ij} relates to T_i in the setup conditional on $s \ni i$ similarly as T_j to T in the original setup.

Finally we may formulate the following basic theorem:

THEOREM 9.9 Under assumptions (9.11) through (9.13), we have

$$\pi_i = 1 - e^{-\alpha_i t} + o(1) \qquad 1 \le i \le N \tag{9.42}$$

where t is a solution of (9.6) or (9.8), and

$$\pi_i\pi_j - \pi_{ij} = (1 - \pi_i)(1 - \pi_j)$$
$$\times \left\{\frac{\pi_i\alpha_j + \pi_j\alpha_i}{\sum_{k=1}^N \alpha_k(1 - \pi_k)} - \alpha_i\alpha_j \frac{\sum_{k=1}^N \pi_k(1 - \pi_k)}{[\sum_{k=1}^N \alpha_k(1 - \pi_k)]^2}\right\}$$
$$\times [1 + o(1)] \tag{9.43}$$

where $o(1) \to 0$ uniformly in i, j, $1 \le i \ne j \le N$.

Proof. The proof follows from (9.15), (9.19), (9.22), (9.23), and (9.41). Q.E.D.

Remark 9.1 As with rejective sampling, the asymptotic expression for $\pi_i\pi_j - \pi_{ij}$ provided by (9.43) can be modified into a tight form without impairing its asymptotic validity by putting

$$\pi_i\pi_j - \pi_{ij} \sim (1 - \lambda_i')(1 - \lambda_j')\left[\frac{\pi_i\alpha_j + \pi_j\alpha_i}{a'} - \alpha_i\alpha_j \frac{d'}{(a')^2}\right] \tag{9.44}$$

where

$$a' = \sum_{k=1}^N \alpha_k(1 - \lambda_k') \qquad d' = \sum_{k=1}^N \pi_k(1 - \lambda_k') \tag{9.45}$$

and the numbers λ_i' satisfy the following implicit relation:

$$(1 - \lambda_i')\left[1 - \frac{\pi_i(1 - \lambda_i')}{d'} + \frac{(1 - \lambda_i')d'}{\pi_i}\left(\frac{\alpha_i}{a'} - \frac{\pi_i}{d'}\right)^2\right] = (1 - \pi_i)$$
$$1 \le i \le N \tag{9.46}$$

As in formula (7.28) the term *tight* means that the sum over all $j \ne i$ is the same on the both sides of (9.44) for every i = 1, ..., N, which is ensured by (9.46). Comparing (9.46) with (3.24), we can see that

$$1 - \lambda_i' \le 1 - \lambda_i \tag{9.47}$$

As an approximation for $1 - \lambda_i'$ we can take

$$1 - \lambda_i' = (1 - \pi_i)\left[1 - \frac{\pi_i(1 - \pi_i)}{d} + \frac{(1 - \pi_i)d}{\pi_i}\left(\frac{\alpha_i}{a} - \frac{\pi_i}{d}\right)^2\right]^{-1} \tag{9.48}$$

where

$$a = \sum_{k=1}^{N} \alpha_k(1 - \pi_k) \qquad d = \sum_{k=1}^{N} \pi_k(1 - \pi_k) \tag{9.49}$$

Remark 9.2 If n is fixed, $N \to \infty$, and $N \max_{1\le i\le N} \pi_i$ remains bounded, (9.44) remains correct up to an error of order N^{-4}. Then, however, we may introduce further simplifications, putting $1 - \lambda_i' = 1 - (n - 1)\pi_i/n$ and $\alpha_i/a' = \pi_i/d'$. Thus, for n fixed, and with precision up to $O(N^{-3})$, there is no difference between rejective and successive sampling as far as $\pi_i\pi_j - \pi_{ij}$ is concerned (see Problem 9.4).

Example 9.3 Table 9.2 shows exact values of $\pi_i\pi_j - \pi_{ij}$ for a successive sampling and three approximations to them. The best fit is provided by formula (9.44) which is asymptotically valid and tight. However, the fit is only slightly better than by approximations based on formulas holding asymptotically for rejective sampling. This illustrates the fact that there is only a minor difference between the values for $\pi_i\pi_j - \pi_{ij}$ for successive and rejective samplings, if they coincide in the π_i terms. The last conclusion is even better supported by the comparison of the first part of Table 9.2 with the first part of Table 7.1, where we may find exact values of $\pi_i\pi_j - \pi_{ij}$ for a rejective sampling whose π_i terms are almost the same as of the successive sampling in Table 9.2. The respective π_i terms are given in Examples 7.3 and 9.1, and also in Table 7.2.

Example 9.4 We shall draw a successive sample of size n = 50 from the population of Table 6.1 with π_i terms proportionate to the x_i terms. For α_i terms we shall use the approximation

TABLE 9.2 Successive sampling with $N = 10$, $n = 4$, $\alpha_i = i/55$, $1 \le i \le 10$; approximations to $\pi_i\pi_j - \pi_{ij}$

i \ j	2	3	4	5	6	7	8	9	10
1	51	70	85	95	102	105	106	104	101
2		128	155	173	185	190	191	187	181
3			210	235	249	256	255	249	240
4				280	296	302	300	291	279
5					326	331	327	316	302
6						345	339	326	310
7							338	324	308
8								315	297
9									283

$10^4(\pi_i\pi_j - \pi_{ij})$ --exact values

i \ j	2	3	4	5	6	7	8	9	10
1	55	74	88	98	103	104	103	101	96
2		135	161	177	186	189	187	182	174
3			216	238	250	254	251	243	233
4				282	296	299	296	286	273
5					324	328	323	313	298
6						342	337	325	310
7							339	327	311
8								320	303
9									291

$10^4 \times$ (9.44) with λ_i' from Eq. (9.48)

i \ j	2	3	4	5	6	7	8	9	10
1	57	76	90	98	103	104	102	99	95
2		138	163	178	186	188	185	179	171
3			219	240	250	253	249	241	230
4				283	296	298	294	285	272
5					324	327	322	312	298
6						342	337	326	311
7							340	329	314
8								324	310
9									300

$10^4 \dfrac{\pi_i(1 - \lambda_i)\pi_j(1 - \lambda_j)}{\sum_{k=1}^{N} \pi_k(1 - \lambda_k)}$ with λ_i from Eq. (3.26)

TABLE 9.2 (continued)

i \ j	2	3	4	5	6	7	8	9	10
1	56	76	90	98	103	103	102	98	94
2		137	162	178	186	187	185	178	171
3			218	240	250	253	249	241	230
4				283	296	299	295	284	298
5					325	328	323	312	306
6						343	338	326	312
7							342	329	315
8								325	310
9									300

$$10^4 \frac{\pi_i(1 - \lambda_i)\pi_j(1 - \lambda_j)}{\sum_{k=1}^{N} \pi_k(1 - \lambda_k)}$$

with λ_i from Eq. (3.24)

$$\begin{aligned} \alpha_i &= \lambda\pi_i\left(1 + \frac{1}{2}\pi_i\right) \\ &= \lambda\left[\frac{50x_i}{4565} + \frac{1}{2}\left(\frac{50x_i}{4565}\right)^2\right] \\ &= \lambda^*\left(100x_i + \frac{2500x_i^2}{4565}\right) \end{aligned}$$

Thus the α_i terms should be proportionate to $100k + .55k^2$ if $x_i = k$. The respective values are given in Table 9.3. We shall perform the sampling by Lahiri's method, thus avoiding the computation of cumulative totals for Table 6.1. A draw of one unit will be realized by drawing randomly a row and, independently, a column, in Table 6.1, and also an auxiliary number ξ from the interval (0,999). If $\xi < 100x_i + .55x_i^2$, we accept the unit, and we repeat the procedure otherwise. If the first digit in ξ is smaller than x_i, then the unit will be accepted irrespective of the other digits; if it is greater than x_i, the draw is void irrespective of the remaining

TABLE 9.3 Values of $100k + .55k^2$, $1 \le k \le 10$

k	1	2	3	4	5	6	7	8	9
$100k + .55k^2$	101	202	305	409	514	620	727	835	944

digits; if the first digit is equal to x_i, then the remaining two digits will be influential. Thus in Table 9.4 we have recorded only the first digit of ξ in all cases where it was not equal to x_i. In Table 9.4 the units accepted in the sample are denoted by A in the fourth column. The unit must be also rejected if it has been included in some previous draw. These units are denoted as a "duplicate." We can see that in selecting only one unit (the last but four draw) the inequality $\xi_i < 100x_i + .55x_i^2$ gave a different decision that $\xi_i < 100x_i$ would give. See also Example 15.3 in this matter.

TABLE 9.4 Successive sample of size n = 50 from the population of Table 6.1, with $\pi_i = 50x_i/4565$, $1 \le i \le 1000$

Row	Column	ξ_i	x_i	Decision	Row	Column	ξ_i	x_i	Decision
1	19	7	3		16	37	5	3	
16	05	2	1		18	40	483	4	
15	15	9	3		9	22	390	3	
17	32	9	2		1	30	8	2	
10	39	126	1		9	18	1	2	A
21	9	0	3	A	3	16	9	6	
3	7	4	5	A	25	17	9	4	
16	15	467	4		11	38	1	4	A
17	14	336	3		3	2	5	2	
19	11	0	7	A	1	3	4	6	A
2	3	2	8	A	16	21	0	3	A
15	10	137	1		14	3	7	1	
15	29	685	6		10	25	3	7	A
8	8	1	2	A	7	20	7	1	
21	15	2	3	A	3	20	563	5	
25	29	1	7	A	12	19	1	6	A
23	35	9	7		5	34	5	2	
22	34	1	5	A	4	11	6	3	
25	24	8	4		13	19	6	5	
7	6	2	4	A	4	26	7	3	

TABLE 9.4 (continued)

Row	Column	ξ_i	x_i	Decision	Row	Column	ξ_i	x_i	Decision
16	21	4	3		24	5	9	8	
12	11	8	2		19	35	4	3	
4	33	6	4		13	2	1	3	A
11	24	3	9	A	19	8	0	5	A
6	40	8	6		8	27	3	5	A
6	30	7	3		4	29	1	8	A
22	14	4	5	A	21	24	6	5	
8	1	899	8		25	5	9	3	
10	31	8	3		3	16	0	6	A
8	1	1	8	A	4	36	7	2	
11	23	1	3	A	3	5	7	8	A
6	10	4	3		9	26	4	2	
1	40	6	9	A	9	30	0	8	A
15	32	4	5	A	13	39	6	9	A
4	29	7	8	A	21	11	7	9	A
8	29	4	5	Duplicate	10	23	6	4	
3	39	2	4	A	4	40	1	7	A
1	16	9	6		21	31	5	7	A
14	34	9	1		11	39	5	2	
4	2	8	4		19	17	1	2	A
16	1	5	4		10	18	4	3	
15	14	1	9	A	4	35	8	5	
2	33	8	4		10	15	7	1	
8	13	3	2		1	29	2	7	A
16	29	8	4		14	35	5	3	
19	31	2	7	A	20	23	1	2	A
7	38	7	1		8	11	4	5	A
4	15	0	9	A	22	6	7	3	
7	6	8	4		12	32	1	2	A
15	30	2	5	A	22	20	4	2	
2	18	3	1		12	39	3	2	

TABLE 9.4 (continued)

Row	Column	ξ_i	x_i	Decision	Row	Column	ξ_i	x_i	Decision
9	15	464	4		6	20	932	9	A
7	23	0	3	A	18	38	2	7	A
4	19	1	5	A	28	1	6	5	
9	33	754	7		8	3	5	6	A
15	9	3	9	A					
20	34	6	8	A					
2	3	6	8	Duplicate					
20	22	8	4						
21	1	6	3						

PROBLEMS AND NOTES

9.1 Perform a successive sampling of size n = 30 from the population of Table 9.1 with probabilities π_i proportionate to the x_i terms.

9.2 Estimate the expected number of draws for the sampling of Problem 9.1.

9.3 *Case n = 2* If n = 2, then

$$\pi_{ij} = \alpha_i\alpha_j\left(\frac{1}{1 - \alpha_i} + \frac{1}{1 - \alpha_j}\right) \tag{9.50}$$

$$\pi_i = \alpha_i\left(1 + \sum_{j \neq i} \frac{\alpha_j}{1 - \alpha_j}\right) \tag{9.51}$$

Show that

$$\pi_i\pi_j \geq \pi_{ij} \geq \alpha_i\alpha_j(1 - \alpha_i - \alpha_j) \geq 0 \tag{9.52}$$

9.4 Assume that n is fixed, $N \to \infty$, and $N \max_{1 \leq i \leq N} \pi_i$ remains bounded. Then

$$\alpha_i = p_i\left[1 + \frac{1}{2}(n - 1)p_i - \frac{1}{2}(n - 1)\sum_{j=1}^{N} p_j^2\right] + O(N^{-3}) \tag{9.53}$$

where $p_i = \pi_i/n$, $1 \le i \le N$. Also the remaining forms of formula (9.9) have the same order of precision. Furthermore,

$$\pi_i\pi_j - \pi_{ij} = \frac{\pi_i\{1 - [(n-1)/n]\pi_i\}\pi_j\{1 - [(n-1)/n]\pi_j\}}{\sum_{k=1}^{N} \pi_k\{1 - [(n-1)/n]\pi_k\}} + O(N^{-4}) \qquad (9.54)$$

(Rao, 1963)

9.5 Under assumptions of Problem 9.4, we have for successive sampling

$$\pi_i\pi_j - \pi_{ij} = \frac{\pi_i(1 - \lambda_i)\pi_j(1 - \lambda_j)}{d} \times \left[1 - \frac{1}{4}\frac{(n-1)(n+5)}{n^2}(\pi_i - \bar{\bar{\pi}})(\pi_j - \bar{\bar{\pi}})\right] + O(N^{-5}) \qquad (9.55)$$

where $d = \Sigma_{i=1}^{N} \pi_i(1 - \lambda_i)$, $1 - \lambda_i$ is given by (3.24) and $\bar{\bar{\pi}} = \Sigma_{i=1}^{N} \pi_i^2(1 - \lambda_i)/d$.

10

SYSTEMATIC SAMPLING

In systematic sampling we must have the units arranged in a sequence. For simplicity we shall assume this arrangement to be such that the unit i is the ith member of the sequence. Then a systematic sample of size n with probabilities of inclusion $\pi_i = n\, x_i/X$ may be drawn as follows:

We realize a random variable ξ having the uniform distribution over the interval $(0,X/n)$ and compute n numbers ξ, $\xi + X/n$, ..., $\xi + (n-1)X/n$, and N cumulative totals x_1, $x_1 + x_2$, ..., $x_1 + \cdots + x_N$. The sample then consists of the units $i_1, \ldots, i_n$ for which

$$x_1 + \cdots + x_{i_{k-1}} < \xi + k\,\frac{X}{n} \le x_1 + \cdots + x_{i_k}$$

$$k = 0, \ldots, n-1 \qquad (10.1)$$

If X/n is an integer, we may realize ξ as a random integer in the range (1,X/n). If X/n is not an integer, we multiply it by a power of 10 and then proceed similarly.

Example 10.1 The following miniature example in Table 10.1 shows a systematic sample of size n = 2 from N = 10. We have X/n = 99/2 = 44.5 and ξ = 36.3. We realized ξ as .1q, where q was a random integer from the range (1,445). The sample is s = {4,8}.

The above procedure selects the unit i at most once, if

$$\max_{1\le i\le N} x_i < \frac{1}{n} X \qquad (10.2)$$

Thus, under (10.2), there will be no duplicates in the sample.

TABLE 10.1 Illustration of systematic sampling with selected units encircled

Unit	x_i	Cumulative totals $x_1 + \cdots + x_i$	ξ / $\xi + \frac{99}{2}$
1	12	12	
2	4	16	
3	17	33	
④	15	48	36.3
5	3	51	
6	10	61	
7	14	75	
⑧	8	83	80.8
9	5	88	
10	11	99	

THEOREM 10.1 Assume that (10.2) holds. In systematic sampling described above the sample s contains a unit i with probability

$$\pi_i = \frac{nx_i}{X} \qquad 1 \le i \le N \tag{10.3}$$

Proof. If for some integer k,

$$(k - 1)\frac{X}{n} \le x_1 + \cdots + x_{i-1} < x_1 + \cdots + x_i \le k\frac{X}{n}$$

holds, then the unit i will be selected if $\xi + (k - 1)X/n$ belongs to the interval $(x_1 + \cdots + x_{i-1}, x_1 + \cdots + x_i)$, which has the probability just (10.3). If, alternatively, for some integer k,

$$x_1 + \cdots + x_{i-1} < (k - 1)\frac{X}{n} < x_1 + \cdots + x_i \le k\frac{X}{n}$$

holds, then the unit i will be selected if either $\xi + (k - 1)X/n$ falls in the interval $((k - 1)X/n, x_1 + \cdots + x_i)$ or if $\xi + (k - 2)X/n$ falls in the interval $(x_1 + \cdots + x_{i-1}, (k - 1)X/n)$. Both events are disjoint and their respective probabilities are

$$\frac{x_1 + \cdots + x_i - (k-1)X/n}{X/n} \quad \text{and} \quad \frac{(k-1)X/n - (x_1 + \cdots + x_{i-1})}{X/n}$$

Their sum again yields (10.3). Q.E.D.

We note that π_i depends only on x_i and on X in a way which is invariant under ordering of units. Thus (10.3) holds for any possible arrangement of the units into a sequence. Especially, if we arrange the units randomly, giving to each permutation the same probability 1/N!, we obtain the so-called *randomized systematic sampling*, suggested first by Madow (1949).

THEOREM 10.2 Equation (10.3) holds also for randomized systematic sampling.

As for the second-order inclusion probabilities, their exact or even asymptotic values are apparently very complicated. As a first approximation we may assume that the π_{ij} terms are for the randomized systematic sampling of the productive type (7.23), i.e., that the sampling can be compared with rejective sampling.

Hartley and Rao (1962) made an attempt to provide an asymptotic theory of randomized systematic sampling for n fixed and $N \to \infty$. Their method is not quite rigorous, because they apply the Edgeworth expansion too freely, and their results, if considered as theorems, are probably not true. This may be demonstrated by the following.

Example 10.2 Consider a population of N units, where N is odd, and a randomized systematic sample of size n = 2 with the following probabilities of inclusion:

$$\pi_1 = \pi_2 = \frac{1}{N-1} \qquad \pi_3 = \cdots = \pi_N = \frac{2}{N-1}$$

Then we shall show that, for any N,

$$\pi_{12} = C$$

which contradicts the Hartley and Rao (1962) Eq. (4.1.26), claimed to be correct to the order $O(N^{-3})$, namely

$$\pi_{12} = \frac{(N - 2)\pi_1\pi_2}{(N - 1)(2 - \pi_1 - \pi_2)} (1 - h^{-2})$$

where h is defined by Eq. (4.1.17) of Hartley and Rao (1962).

Actually, applying our definition of randomized systematic sampling with $\pi_i \equiv x_i$, the pair (1,2) will be selected with a positive probability, if there is a number $\xi \in (0,1)$ and an integer k such that

$$0 < \xi \leq \frac{1}{N - 1} \tag{10.4}$$

$$\frac{1}{N - 1} + \frac{2k}{N - 1} < \xi + 1 \leq \frac{(k + 1)2}{N - 1} \tag{10.5}$$

Here, without losing generality, we have assumed that the unit 1 was chosen first in the random arrangement, and that between the units 1 and 2 there are k units from among 3, 4, ..., N. The inequality (10.5) may be rewritten as

$$\frac{2k + 2 - N}{N - 1} < \xi \leq \frac{2k + 3 - N}{N - 1} \tag{10.6}$$

Now the inequalities (10.4) and (10.6) are incompatible for any integer k if N is odd. For example, if $k = (1/2)(N - 1)$, then (10.6) becomes

$$\frac{1}{N - 1} < \xi \leq \frac{2}{N - 1}$$

and if $k = (1/2)(N - 3)$, then (10.5) becomes

$$-\frac{1}{N - 1} < \xi \leq 0$$

Thus there is hardly a hope for reasonable asymptotic expressions for π_{ij} for $n = 2$ and $N \to \infty$. The Hartley and Rao results could apparently be given some interpretation if $\pi_1, \ldots, \pi_N$ might be regarded as a sample from a distribution possessing a sufficiently smooth density.

Remark 10.1 In Chapters 7 and 9 we have mentioned selection of one unit with drawing probabilities α_i by means of cumulative totals. This may be considered as systematic sampling of size $n = 1$ and can be illustrated by means of Table 10.1. Putting $\alpha_i = x_i/X$, the sampling is carried out by means of a random number ξ from the interval $(1,X)$, while the selected unit i satisfies

$$x_1 + \cdots + x_{i-1} < \xi \leq x_1 + \cdots + x_i$$

For example, if $\xi = 24$, then $x_1 + x_2 = 16 < \xi \leq 33 = x_1 + x_2 + x_3$ and the selected unit is $i = 3$. Also, Table 5.1 could be used as an illustration if the units are identified with counties and x_i is the number of schools in the county.

PROBLEMS AND NOTES

10.1 Take a uniform ($\pi_i = n/N$) systematic sample of size $n = 100$ from the population of Table 6.1, assuming that the units are ordered primarily by rows and secondarily by columns. Compare the sample frequencies of the digits 1, ..., 9 with expected frequencies.

10.2 Take a systematic sample of size $n = 100$ with inclusion probabilities π_i proportionate to the digits x_i, assuming the same ordering as in Problem 10.1. Again, compare the sample frequencies of the digits 1, ..., 9 with expected frequencies.

11

STRATIFIED SAMPLING

Very often we have reason to consider the population as divided into a number of disjoint subpopulations. In sample surveys these subpopulations are called *strata*. We shall denote the number of strata by L and the individual strata by S_h, h = 1, 2, ..., L. Thus

$$S = \bigcup_{h=1}^{L} S_h \tag{11.1}$$

The subdivision of S into strata may be motivated either by the fact that we dispose of a separate frame for each stratum, or that we expect within strata less variability of the characteristic under consideration than in the population on the whole, or, finally, different strata may correspond to different cost levels.

The simplest way to treat a stratified population is to apply a sampling method in each stratum *independently*. For example, we may apply rejective sampling in each stratum with inclusion probabilities proportionate to some numbers x_i. Though the population is stratified, we shall index the units by simple indices i = 1, 2, ..., N in order to avoid confusion in the notation applied to first- and second-order inclusion probabilities π_i and π_{ij}, respectively. Let h(i) be the index of the stratum to which the unit i belongs. Thus h(i) = h means that $i \in S_h$.

If we denote the stratum total by

$$X_h = \sum_{i \in S_h} x_i \tag{11.2}$$

and apply formula (3.31) to individual strata, we obtain

$$\pi_i = n_h \frac{x_i}{X_h} \qquad i = 1, 2, \ldots, N \qquad h = h(i) \tag{11.3}$$

where n_h are sample sizes in the strata. Note that the total sample size equals

$$n = \sum_{h=1}^{L} n_h \tag{11.4}$$

and the total of the x_i terms equals

$$X = \sum_{h=1}^{L} X_h \tag{11.5}$$

In order to achieve the π_i terms proportionate to the x_i terms throughout the population, the ratios n_h/X_h should equal a constant λ, and it is easy to see that then $\lambda = n/X$. Thus, if

$$n_h = X_h \frac{n}{X} \qquad h = 1, \ldots, L \tag{11.6}$$

the equation

$$\pi_i = \frac{nx_i}{X} \tag{11.7}$$

holds even with stratified sampling.

If the inclusion probabilities are uniform, i.e., $x_i \equiv 1$, then (11.6) and (11.7) become

$$n_h = N_h \frac{n}{N} \qquad h = 1, \ldots, L \tag{11.8}$$

and

$$\pi_i = \frac{n}{N} \qquad i = 1, \ldots, N \tag{11.9}$$

respectively, where N_h denotes the size of the stratum S_h, and $N = \Sigma_{h=1}^{L} N_h$.

In order to be able to approximate reasonably the n_h terms in (11.6) by integers, the strata must be either large enough or created in such a way that the right sides of (11.6) are integers, or closely so.

If the sampling in each stratum is carried out independently, the indicators I_i and I_j will be independent, provided that i and j do not belong to the same stratum. Thus

$$\pi_i\pi_j - \pi_{ij} = 0 \qquad \text{if } h(i) \neq h(j) \qquad i,j = 1, \ldots, N \qquad (11.10)$$

On the other hand, if i and j are members of the same stratum, formula (7.27) is applicable, and we have for rejective sampling

$$\pi_i\pi_j - \pi_{ij} \sim \frac{\pi_i(1 - \pi_i)\pi_j(1 - \pi_j)}{d_h} \qquad h(i) = h(j) \qquad (11.11)$$

where

$$d_h = \sum_{i \in S_h} \pi_i(1 - \pi_i)$$

Since $d_h < d = \Sigma_{i=1}^N \pi_i(1 - \pi_i) = \Sigma_{h=1}^L d_h$, the effect of stratification on correlations of inclusion indicators is that they are nullified between strata and increased (in absolute value) within strata. On the average the correlation within strata is L times larger than for sampling without stratification.

Remark 11.1 If we apply Poisson sampling, there is no difference between stratified and plain sampling, since each unit is treated independently in either case.

Example 11.1 If sampling a population consisting of persons, we may form strata according to territorial aspect, rural-urban aspect, profession, sex, age, or income.

Example 11.2 Assume that the population of Table 6.1 splits into four strata as follows:

S_1 ... rows 1 through 5
S_2 ... rows 6 through 15
S_3 ... rows 16 through 18
S_4 ... rows 19 through 25

We wish to perform sampling of size $n = 100$ with π_i terms proportionate to the x_i terms in each stratum, and choose the sample sizes $n_1, \ldots, n_4$ in the strata in such a manner that (11.7) holds throughout the whole population. Consequently we have to apply formula (11.6). The strata totals were established with the following result:

$$X_1 = 912 \qquad X_2 = 1830 \qquad X_3 = 536 \qquad X_4 = 1287$$

Thus

$$n_1 = \frac{912 \times 100}{4565} = 19.98 \doteq 20 \qquad n_3 = \frac{536 \times 100}{4565} = 11.74 \doteq 12$$

$$n_2 = \frac{1830 \times 100}{4565} = 40.09 \doteq 40 \qquad n_4 = \frac{1287 \times 100}{4565} = 28.19 \doteq 28$$

PROBLEMS AND NOTES

11.1 Split the population of Table 6.1 into three strata as follows: S_1--colums 1 through 10, S_2--columns 11 through 25, S_3--columns 26 through 40. Then allocate a stratified sample of total size $n = 100$ if we wish the π_i terms to be

(a) Uniform ($\pi_i = n/N$)

(b) Proportionate to the x_i terms

throughout the whole population.

11.2 Take a stratified simple random sample according to Problem 11.1(a) and a stratified successive sample according to Problem 11.1(b).

11.3 Find the "tight" form, corresponding to (7.28), for Eq. (11.11).

12

MULTISTAGE SAMPLING

If the population is very large, we prefer to carry out the sampling in *several stages*. For the first stage we define a new population whose units are clusters of the original units. The clusters used for the first stage of sampling are called *primary sampling units* (psu).

Example 12.1 If sampling a population of persons the psu may be the houses. If sampling schoolchildren the psu may be schools.

Each primary sampling unit selected in the first stage may be considered as a smaller population from which we select a certain number of smaller clusters--*secondary sampling units* (ssu). Then we may stop or select from the included ssu a certain number of *tertiary sampling units* (tsu). The units used in the last stage are called *ultimate units*. They may be either units possessing the characteristics at issue (*elementary units*) or some small clusters of them. For example, in the population of inhabitants the ultimate units may be considered either as persons or as families. Secondary, tertiary, etc., stages are called *subsampling*.

Unless stated otherwise, the second-stage sampling in each psu is carried out independently, and a similar assumption applies to further stages.

If a unit i belongs to the rth psu, $r = r(i)$, then the unit will be included in the sample s if

1. the rth psu will be included in the first-stage sample; and
2. if the unit i will be selected in subsampling, provided that case 1 happened.

Denoting the probability of case 1 by π_r^I and the conditional probability in 2 by π_i^{II}, we have the following formula for the overall inclusion probability:

$$\pi_i = \pi_r^I \pi_i^{II} \qquad i = 1, \ldots, N \qquad r = r(i) \tag{12.1}$$

Example 12.2 Let us have M psu and apply simple random sampling of size m in the first stage. Furthermore, let the rth psu consist of N_r elementary units out of which we select n_r by simple random sampling in the second stage, if any. Then

$$\pi_i = \frac{m}{M}\frac{n_r}{N_r} \qquad r = 1, \ldots, M \qquad i = 1, \ldots, N \qquad r = r(i) \tag{12.2}$$

Thus the sampling will be overall uniform if

$$n_r = N_r \frac{n}{N} \qquad \text{where } n = \sum_{r=1}^{M} n_r \text{ and } N = \sum_{r=1}^{M} N_r \tag{12.3}$$

Example 12.3 If we have M psu and the first stage is rejective sampling of size m with probabilities of inclusion proportionate to x_r terms, and the second state is the same as in Example 12.2, then

$$\pi_i = \frac{mx_r}{X}\frac{n_r}{N_r} \qquad i = 1, \ldots, N \qquad r = 1, \ldots, M \qquad r = r(i) \tag{12.4}$$

where $X = \Sigma_{r=1}^M x_r$. Thus if we want uniformity in the π_i terms we must have

$$n_r = \frac{\pi X N_r}{m x_r} \qquad r = 1, \ldots, M \tag{12.5}$$

where π is the overall value of π_i.

Now we shall consider second-order inclusion probabilities in multistage sampling. Again, for two units i and j belonging to rth and sth psu, respectively, we may write

$$\pi_{ij} = \pi_{rs}^I \pi_{ij}^{II} \qquad i,j = 1, \ldots, N \qquad r = r(i) \qquad s = s(j) \tag{12.6}$$

where π^I_{rs} denotes the probability of simultaneous inclusion of the rth and sth psu in the first stage, and π^{II}_{ij} denotes the conditional probability of simultaneous selection of the units i and j in subsampling, provided that the rth and sth psu have been selected in the first stage.

Now, if $r \neq s$, the subsampling concerning the unit i is independent from one concerning the unit j. Thus $\pi^{II}_{ij} = \pi^{II}_i \pi^{II}_j$ and

$$\pi_{ij} = \pi^I_{rs}\pi^{II}_i\pi^{II}_j \qquad \text{if } r(i) \neq s(j) \tag{12.7}$$

If $r = s$, we obviously have $\pi^I_{rs} = \pi^I_r$ and

$$\pi_{ij} = \pi^I_r\pi^{II}_{ij} \qquad \text{if } r(i) = r(j) \tag{12.8}$$

THEOREM 12.1 In multistage sampling as described above we have

$$\pi_i\pi_j - \pi_{ij} = \begin{cases} \pi^{II}_i\pi^{II}_j(\pi^I_r\pi^I_s - \pi^I_{rs}) & \text{if } r(i) \neq s(j) \\ \pi^{II}_i\pi^{II}_j\pi^I_r(\pi^I_r - 1) + \pi^I_r(\pi^{II}_i\pi^{II}_j - \pi^{II}_{ij}) & \\ & \text{if } r(i) = r(j) \end{cases} \tag{12.9}$$

Example 12.4 Under conditions of Example 12.2 we have by (5.6)

$$\pi_i\pi_j - \pi_{ij} = \begin{cases} \dfrac{n_r n_s}{N_r N_s}\dfrac{m(M - m)}{M^2(M - 1)} & \text{if } r(i) \neq s(j) \\ \left(\dfrac{n_r}{N_r}\right)^2 \dfrac{m}{M}\left(\dfrac{m}{M} - 1\right) + \dfrac{m}{M}\dfrac{n_r(N_r - n_r)}{N_r^2(N_r - 1)} & \text{if } r(i) = r(j) \end{cases}$$

Example 12.5 If sampling is rejective of size m in the first stage and rejective of size n_r in the second stage, M and N_r keeping their meaning, we have approximately, by (7.23),

$$\pi_i\pi_j - \pi_{ij} \sim \frac{\pi^{II}_i\pi^{II}_j\pi^I_r(1 - \pi^I_r)\pi^I_s(1 - \pi^I_s)}{d^I} \qquad \text{if } r(i) \neq s(j) \tag{12.10a}$$

$$\pi_i\pi_j - \pi_{ij} \sim \pi_i^{II}\pi_j^{II}\pi_r^{I}(\pi_r^{I} - 1) + \frac{\pi_r^{I}\pi_i^{II}\pi_j^{II}(1 - \pi_i^{II})(1 - \pi_j^{II})}{d_r^{II}}$$

$$\text{if } r(i) = r(j) \qquad (12.10b)$$

where

$$d^{I} = \sum_{r=1}^{M} \pi_r^{I}(1 - \pi_r^{I})$$

and

$$d_r^{II} = \sum_{r(i)=r} \pi_i^{II}(1 - \pi_i^{II})$$

PROBLEM

12.1 Consider the columns in Table 6.1 as psu. Draw a sample of $m = 10$ psu by simple random sampling. Then, within each selected psu, draw a simple random sample of size $n_r = 5$. Establish the overall probabilities of inclusion.

13

ASSORTED METHODS

RATIO VARIANTS OF DESIGNS Consider an arbitrary sampling design P(s) with inclusion probabilities $\pi_1, \ldots, \pi_N$. Further assume that for each unit there is given a measure of "size" $x_i \geq 0$, $1 \leq i \leq N$, and introduce the statistic

$$\hat{X} = \sum_{i \in s} \frac{x_i}{\pi_i} = \sum_{i=1}^{N} \frac{x_i}{\pi_i} I_i \tag{13.1}$$

where I_i are the indicators of inclusion. In Chapter 17 we shall call $\hat{X}$ the simple linear estimator and show that

$$E_P\hat{X} = X \tag{13.2}$$

where E_P is the expectation referring to P(s), and $X = \Sigma_{i=1}^{N} x_i$.

Now, with every design P(s) we may associate a new design Q(s) as follows:

$$Q(s) = \frac{\hat{X}}{X} P(s) \qquad s \subset S \tag{13.3}$$

Relation (13.2) entails that

$$\Sigma_s Q(s) = 1$$

We shall call Q(s) the *ratio variant* of P(s), indicating that this design is useful in connection with ratio estimates (see Chapter 18). The value of Q(s) lies in the following:

THEOREM 13.1 The ratio variant Q(s) satisfies

$$E_Q\left(\frac{X}{\hat{X}} I_i\right) = \pi_i \qquad 1 \le i \le N \tag{13.4}$$

where E_Q is the expectation referring to Q(s).

Proof. Equation (13.3) implies that

$$E_Q\left(\frac{X}{\hat{X}} I_i\right) = E_P(I_i) = \pi_i \qquad 1 \le i \le N$$

which concludes the proof.

Ratio variant of simple random sampling If P(s) is given by (5.1), then (13.3) yields

$$Q(s) = \begin{cases} \frac{x}{X}\binom{N-1}{n-1}^{-1} & \text{if size } s = n \\ 0 & \text{otherwise} \end{cases} \tag{13.5}$$

where $x = \Sigma_{i \in s}\, x_i$.

This sampling was first suggested by Hájek (1949) under the label "two-phase sampling," although priority is usually ascribed to Lahiri (1951) and Midzuno (1951). The procedure may be practically realized in two steps:

1. We draw one unit with drawing probabilities proportionate to x_i, $\alpha_i = x_i/X$, $1 \le i \le N$.

2. Then we select a simple random sample of size n - 1 from the remaining N - 1 units.

The sample then consists of all n units obtained in steps 1 and 2.

Ratio variant of rejective sampling If P(s) is given by (7.2), we may realize the ratio variant Q(s) again in two steps:

1. We draw a unit with single-draw probabilities $\alpha_i^I = \lambda x_i \alpha_i/\pi_i$, $1 \le i \le N$, where α_i are single-draw probabilities for the corresponding rejective sampling.

2. Then we select a rejective sample of size n - 1 with single-draw probabilities α_i from the whole population.

If all the n units drawn in steps 1 and 2 are distinct, we accept the sample; otherwise we reject it as soon as the first duplicate appears and start again from step 1.

Ratio variant of an arbitrary sampling For successive sampling and for any other sampling we may proceed as follows:

1. We realize a sample s in accordance with probabilities P(s).

2. We compute $\hat{X} = \hat{X}(s)$ and choose a number K such that $\hat{X} \le K$ holds always or with a probability close to 1. Then we select a random number ξ uniformly distributed in the range (0,K) and accept the sample if $\xi \le \hat{X}$; otherwise we reject it and start again with step 1.

PERMUTATION SAMPLING Permutation sampling of average size n with probabilities of inclusion $\pi_1, \ldots, \pi_N$ is performed as follows: We arrange the units randomly into a sequence $(i_1, \ldots, i_N)$ and the sample consists of units i_k such that

$$k + \xi_k \le N\pi_{i_k} \tag{13.6}$$

where $\xi_1, \ldots, \xi_N$ are independent random variables distributed uniformly over (0,1). If we see that ξ_k cannot matter in inequality (13.6), we may ignore it. This will be possible for the majority of k terms, and, if all $N\pi_i$ are integers, it will be possible for all k terms. If the π_i terms are proportionate to some x_i terms, we set $\pi_i = nx_i/X$.

THEOREM 13.2 In the permutation sampling described above the π_i terms used in (13.6) are the probabilities of inclusion, and if the numbers $N\pi_i$ are integers, then

$$\pi_i\pi_j - \pi_{ij} = \frac{\pi_i(1 - \pi_j)}{N - 1} \qquad \text{if } \pi_i \le \pi_j \tag{13.7}$$

Proof. The proof is left to the reader as an exercise.

Permutation sampling may be regarded as a modification of Poisson sampling aiming at the reduction of the variability of the sample size. If $\pi_i = n/N$, $1 \le i \le N$, the permutation sampling coincides with simple random sampling and Var(size s) = 0. If the π_i terms are variable, then Var(size s) > 0, but it will be only a a fraction of the variance under Poisson sampling (say, one-quarter at most).

One may ask what sampling we obtain if permutation sampling is conditioned by size s = n. Assuming that conditional covariance of I_i and I_j may be approximated by residual covariance under linear regression of I_i and I_j on K = size s, we obtain

$$\pi_i\pi_j - \pi_{ij} \sim \frac{\pi_i(1 - \pi_j)}{N - 1} + \frac{C_iC_j}{C} \qquad \pi_i \le \pi_j \tag{13.8}$$

where

$$C_i = \pi_i(1 - \pi_i) - \frac{\sum_{j\neq i}^{N} \min(\pi_i,\pi_j)[1 - \max(\pi_i,\pi_j)]}{N - 1} \tag{13.9}$$

and

$$C = \sum_{i=1}^{N} C_i \tag{13.10}$$

This suggests that we obtain a sampling which is different from rejective sampling even asymptotically, unless $\pi_i = n/N$, $1 \le i \le N$ (cf. Problem 7.7).

Permutation sampling was suggested by Hájek (1959).

RANDOM GROUPS SAMPLING This sampling was proposed by Cochran et al. (1962) and consists of two stages:

1. Split the population *at random* in n groups of sizes N_1, N_2, ..., N_n, where $N_1 + N_2 + \cdots + N_n = N$.

2. Draw a sample of size 1 with probabilities proportionate to x_i from each of these groups independently.

Let X_i denote the sum of the x values over the group to which the unit i belongs. Obviously, $X_i = X_j$ if the units i and j belong to the same group. The X_i terms are random variables depending on random grouping.

The inclusion probabilities equal, obviously,

$$\pi_i = x_i E\left(\frac{1}{X_i}\right) \qquad 1 \le i \le N \tag{13.11}$$

Given a grouping G, we have

$$E(I_iI_j|G) = \begin{cases} x_ix_j/X_iX_j & \text{if i and j do not belong to the same group} \\ 0 & \text{otherwise} \end{cases} \tag{13.12}$$

and

$$\pi_{ij} = E(E(I_iI_j|G)) \tag{13.13}$$

where the expectation E extends over the groupings G.

In order to reduce the variability of the X_i terms, it is advisable to have the N_j terms--sizes of the groups--as uniform as possible.

If the sampling fraction n/N is not very large, the random groups sampling will be very close to rejective sampling.

PROBLEMS AND NOTES

13.1 Show that for the ratio variant of simple random sampling the following holds:

$$\pi_i = \frac{n-1}{N-1} + \frac{N-n}{N-1}\frac{x_i}{X} \qquad 1 \le i \le N \tag{13.14}$$

13.2 Show that for the same sampling we have:

$$\pi_i\pi_j - \pi_{ij} = \frac{(n-1)(N-n)}{(N-1)^2(N-2)} - \frac{(n-1)(N-n)}{(N-1)^2(N-2)}\frac{x_i + x_j}{X} + \left(\frac{N-n}{N-1}\right)^2 \frac{x_ix_j}{X^2} \qquad 1 \le i \ne j \le N \tag{13.15}$$

13.3 Verify that the procedures of realization of the ratio variant of rejective sampling and of a general sampling given above are consistent with Eq. (13.3).

13.4 Take a permutation sample of size n = 30 with π_i terms proportionate to the x_i terms from the population of Table 6.1. [*Hint:* It is sufficient to allocate the first 60 members of the permutation $(i_1,\ldots,i_N)$, i.e., to draw an *ordered* simple random sample of size n = 60, and then to apply (13.6).]

13.5 Prove Theorem 13.2. Show that (13.7) holds even if $N\pi_i$ and $N\pi_j$ are not integers, and $[N\pi_i] \neq [N\pi_j]$, where $[x]$ denotes the integral part of x.

13.6 Prove that the right-hand side of (13.8) equals the residual covariance of I_i and I_j on K = size s.

14

CONDITIONAL POISSON SAMPLING

Generally we may condition a Poisson sampling by the requirement that the sample s belongs to a subclass Ω of samples. Any such design was shown (Theorem 3.4) to maximize the entropy among the procedures having the same π_i terms and the same carrier.

Many of the preceding sampling methods may be described as conditional Poisson sampling: simple random sampling, stratified or not, rejective sampling, stratified or not, two-stage samplings of Examples 12.2 and 12.3, etc.

An important class of conditional Poisson samplings may be obtained by means of a set of integral-valued random variables $K_j = \text{size}(s \cap S_j)$ where $S_j \subset S$ is a stratum. The conditioning is defined by fixing $K_j = k_j$, $1 \le j \le m$, where, usually, $k_j = EK_j$.

Example 14.1 If $m = 1$ and $K_1 = K = \text{size } s$, we obtain rejective sampling. If $m = L$ and $K_h = \text{size } s_h = \text{size}(s \cap S_h)$, where s_h is the sample from the hth stratum S_h, we obtain stratified rejective sampling if the strata are disjoint.

The basic problem connected with conditioning is to reevaluate the inclusion probabilities. One way is to derive a refined local limit theorem for the random variables K_j and then obtain the conditional inclusion probabilities by the Bayes theorem. This approach was taken in Chapter 7. Alternatively, we may replace conditioning by regression and hope for similarly good approximations. In some cases we get equivalent approximations, as is illustrated by the following:

Example 14.2 Denoting K = size s, again the linear regression predictor of I_i by K equals

$$\hat{\pi}_i = p_i + \frac{p_i(1 - p_i)}{d} (K - n) \qquad 1 \le i \le N \tag{14.1}$$

where $n = EK = p_1 + \cdots + p_N$, and $d = \Sigma_{i=1}^N p_i(1 - p_i)$. If n is an integer, and we condition on K = n, the predictor becomes

$$\hat{\pi}_i = p_i \qquad 1 \le i \le N$$

A more precise result is yielded by the quadratic regression. Easy computations based on (6.10) provide the following quadratic regression predictor for I_i:

$$\hat{\pi}_{iQ} = p_i + \frac{p_i(1 - p_i)}{d} (K - n) + \frac{(K - n)^2 - (1 - 2\bar{\bar{p}})(K - n) - d}{d^2[1 + o(1)]} \times p_i(1 - p_i)(\bar{\bar{p}} - p_i) \tag{14.2}$$

where $d = \Sigma_{i=1}^N p_i(1 - p_i)$ and $\bar{\bar{p}} = \Sigma_{i=1}^N p_i^2(1 - p_i)/d$. Setting $K = n$, we obtain

$$\hat{\pi}_i = p_i\left[1 - \frac{(\bar{\bar{p}} - p_i)(1 - p_i)}{d} + o(d^{-1})\right]$$

which coincides with (7.16). Generally, we could prove that polynomial regression of degree m provides the same values for the π_i terms as the m-term Edgeworth expansion for $P(K = k)$.

Conditional covariances of I_i and I_j are defined by

$$E[(I_i - \tilde{\pi}_i)(I_j - \tilde{\pi}_j) | K_j = k_j,\ 1 \le j \le m] \tag{14.3}$$

where $\tilde{\pi}_i = E(I_i | K_j = k_j,\ 1 \le j \le m)$. The first-order approximation for these values may be obtained by replacing the $\tilde{\pi}_i$ values by regression predictors $\hat{\pi}_i$, and by assuming that the $(I_i - \hat{\pi}_i)(I_j - \hat{\pi}_j)$ and $(K_1,\ldots,K_m)$ are independent. Then (14.3) will be approximated by

$$(14.3) \doteq E[(I_i - \hat{\pi}_i)(I_j - \hat{\pi}_j)]$$

$$= \text{Cov}(I_i, I_j) - \text{Cov}(\hat{\pi}_i, \hat{\pi}_j) \qquad (14.4)$$

Example 14.3 Utilizing linear regression predictors (14.1) and conditioning by $K = n$, we obtain from (14.4) the following approximate covariances of I_i, I_j in rejective sampling:

$$\text{Cov}(I_i, I_j | K = n) \doteq \text{Cov}(I_i, I_j) - \text{Cov}\left[\frac{p_i(1 - p_i)}{d} K, \frac{p_j(1 - p_j)}{d} K\right]$$

$$= -\frac{p_i(1 - p_i)p_j(1 - p_j)}{d} \qquad (14.5)$$

We have used the fact that $\text{Cov}(I_i, I_j) = 0$ and $\text{Var}\ K = d$ under Poisson sampling. We can see from (7.23) and (7.16) that (14.5) is asymptotically valid for $d \to \infty$.

Remark 14.1 Replacement of conditioning by regression is applicable not only to Poisson sampling, but also to other designs, as we have seen in Chapter 13 in connection with permutation sampling. There, however, we were not able to prove the asymptotic equivalence of conditioning and regression.

Let us now show how regression may be employed to establish $\pi_i\pi_j - \pi_{ij}$ to the order $O(1/d^2)$. We shall make use of the fact that, to that order, the residual covariance and expected conditional covariance coincide if we utilize the quadratic regressor (14.2). Thus

$$E\ \text{Cov}(I_i, I_j | K) \sim -\text{Cov}(\hat{\pi}_{iQ}, \hat{\pi}_{jQ}) \qquad (14.6)$$

Obviously (14.2) entails

$$\text{Cov}(\hat{\pi}_{iQ}, \hat{\pi}_{jQ}) \sim \frac{p_i(1 - p_i)p_j(1 - p_j)}{d}\left[1 + 2\frac{(\bar{\bar{p}} - p_i)(\bar{\bar{p}} - p_j)}{d}\right] \qquad (14.7)$$

where $d = \Sigma_{i=1}^N p_i(1 - p_i)$, $\bar{\bar{p}} = \Sigma_{i=1}^N p_i^2(1 - p_i)/d$, and the difference is of order $o(1/d^2)$.

Next we may write

$$\mathrm{Cov}(I_i, I_j | K) \sim \frac{\hat{\pi}_{iQ}(1 - \hat{\pi}_{iQ})\hat{\pi}_{jQ}(1 - \hat{\pi}_{jQ})}{\hat{d}} + \frac{c_{ij}}{d^2} \qquad 1 \le i \ne j \le N \tag{14.8}$$

where the $\hat{\pi}_{iQ}$ terms are given by (14.2), $\hat{d} = \Sigma_{i=1}^{N} \hat{\pi}_{iQ}(1 - \hat{\pi}_{iQ})$, and c_{ij} may be assumed to be independent of K. The independence of c_{ij} of K is only a conjecture, but a very plausible one, based on the assumption that c_{ij} depends on the π_i terms continuously. Once more employing (14.2), we have

$$\frac{\hat{\pi}_{iQ}(1 - \hat{\pi}_{iQ})\hat{\pi}_{jQ}(1 - \hat{\pi}_{jQ})}{\hat{d}} \sim \frac{p_i(1 - p_i)p_j(1 - p_j)}{d} \left\{1 + \xi_{ij} + \frac{[4(\bar{\bar{p}} - p_i)(\bar{\bar{p}} - p_j) - p_i(1 - p_i) - p_j(1 - p_j) + \overline{p(1 - p)}](K - n)^2}{d^2}\right\} \tag{14.9}$$

where $\overline{p(1 - p)} = \Sigma_{i=1}^{N} p_i^2(1 - p_i)^2/d$ and ξ_{ij} is a random variable with $E\xi_{ij} = 0$. Substituting (14.9) into (14.8), we obtain

$$E\,\mathrm{Cov}(I_i, I_j | K) \sim \frac{p_i(1 - p_i)p_j(1 - p_j)}{d} \times \left[1 + \frac{4(\bar{\bar{p}} - p_i)(\bar{\bar{p}} - p_j) - p_i(1 - p_i) - p_j(1 - p_j) + \overline{p(1 - p)}}{d}\right] + \frac{c_{ij}}{d^2} \qquad 1 \le i \ne j \le N \tag{14.10}$$

Identifying the right-hand sides of (14.10) and (14.7) we get

$$c_{ij} \sim p_i(1 - p_i)p_j(1 - p_j)[-2(\bar{\bar{p}} - p_i)(\bar{\bar{p}} - p_j) + p_i(1 - p_i) + p_j(1 - p_j) - \overline{p(1 - p)}] \qquad 1 \le i \ne j \le N \tag{14.11}$$

Thus, for rejective sampling of size n,

$$\pi_i\pi_j - \pi_{ij} = \mathrm{Cov}(I_i, I_j | K = n) \sim \frac{\pi_i(1 - \pi_i)\pi_j(1 - \pi_j)}{d_*} + \frac{c_{ij}}{d_*^2} \tag{14.12}$$

where $d_* = \Sigma_{i=1}^N \pi_i(1 - \pi_i)$. Putting $d_\lambda = \Sigma_{i=1}^N \pi_i(1 - \lambda_i)$ and utilizing (3.24), we have further

$$1 - \pi_i = (1 - \lambda_i)\left[1 - \frac{\pi_i(1 - \lambda_i)}{d_\lambda}\right] \sim (1 - \lambda_i)\left[1 - \frac{p_i(1 - p_i)}{d}\right]$$

Consequently, we may simplify (14.12) as follows:

$$\pi_i\pi_j - \pi_{ij} \sim \frac{\pi_i(1 - \lambda_i)\pi_j(1 - \lambda_j)}{d_\lambda}\left[1 - 2\,\frac{(\bar{\bar{\pi}} - \pi_i)(\bar{\bar{\pi}} - \pi_j)}{d_\lambda}\right] \qquad 1 \le i \ne j \le N \qquad (14.13)$$

where $\bar{\bar{\pi}} = \Sigma_{i=1}^N \pi_i^2(1 - \lambda_i)/d_\lambda$.

Example 14.4 Equation (14.13) provides the value of $\pi_i\pi_j - \pi_{ij}$ with high accuracy. If applied to the data of Example 7.3, we obtain Table 14.1. We can see that the approximations are very close to the rounded-off exact values given in the top part of Table 7.1.

Consider now the statistics

$$\hat{X} = \sum_{i \in s} \frac{x_i}{p_i} = \sum_{i=1}^{N} \frac{x_i}{p_i} I_i \qquad (14.14)$$

TABLE 14.1 Equation (14.13) applied to data of Example 7.3

i \ j	2	3	4	5	6	7	8	9	10
1	52	72	86	97	103	106	106	103	100
2		132	158	176	186	191	190	185	179
3			214	237	250	255	253	247	237
4				281	296	300	297	289	277
5					324	328	323	314	300
6						341	336	325	310
7	$10^4(\pi_i\pi_j - \pi_{ij})$						337	325	310
8	approximated by							318	303
9	Eq. (14.13)								291

where $x_1, \ldots, x_N$ are arbitrary numbers. Obviously, they subsume the statistics K_j as special cases. Assuming that

$$\sum_{i=1}^{N} p_i(1 - p_i) \sum_{i=1}^{N} \frac{x_i^4}{p_i^4} p_i(1 - p_i) \le C \sum_{i=1}^{N} \frac{x_i^2}{p_i^2} p_i(1 - p_i) \qquad (14.15)$$

where C is a fixed constant, we may easily derive that the quadratic regression for I_i in terms of X equals

$$\begin{aligned}\pi_{iQ} = p_i &+ \frac{x_i(1 - p_i)}{D}(\hat{X} - X) \\ &+ \frac{(\hat{X} - X)^2 - \mu_3(\hat{X} - X)/D - D}{2D^2[1 + o(1)]} x_i(1 - p_i)\left[\frac{x_i}{p_i}(1 - 2p_i) \right.\\ &\left. - \frac{\mu_3}{D}\right]\end{aligned} \qquad (14.16)$$

where $X = \Sigma_{i=1}^{N} x_i = E\hat{X}$,

$$D = E(\hat{X} - X)^2 = \sum_{i=1}^{N} \frac{x_i^2}{p_i}(1 - p_i) \qquad (14.17)$$

and

$$\mu_3 = E(\hat{X} - X)^3 = \sum_{i=1}^{N} \frac{x_i^3}{p_i^2}(1 - p_i)(1 - 2p_i) \qquad (14.18)$$

The term $\mu_3(\hat{X} - X)/D$ and $o(1)$ may be omitted in (14.17), since it contributes $o(1/d^2)$.

The linear regression predictor of I_i on the basis of several statistics $\hat{X}_j = \Sigma_{i \in s} x_{ji}/p_i$, $1 \le j \le m$, equals

$$\hat{\pi}_i = p_i + (1 - p_i)x_i D^{-1}(\hat{X} - X)' \qquad 1 \le i \le N \qquad (14.19)$$

where $x_i = (x_{1i}, \ldots, x_{mi})$, and D^{-1} is the inverse of the variance matrix D of the vector $\hat{X} - X = (\hat{X}_1 - X_1, \ldots, \hat{X}_m - X_m)$, $X_j = \Sigma_{i=1}^{N} x_{ji}$, $1 \le i \le m$. If D is singular, D^{-1} denotes one of the generalized inverses of D (see Rao, 1965). If x is a row vector, then x' denotes the corresponding column vector.

The quadratic regression predictor in several variables is a little tedious. Utilizing the fact that in large samples the vector $\hat{X} - X$ behaves similarly to a normal vector, if the excess-boundedness condition (14.15) is satisfied for all its linear functions, we can derive after some computations

$$\begin{aligned}\hat{\pi}_{iQ} &= p_i + (1 - p_i)x_i D^{-1}(\hat{X} - X)' \\ &+ \frac{1}{2}(1 - p_i)\left\langle \frac{1 - 2p_i}{p_i} \{[x_i D^{-1}(\hat{X} - X)']^2 - x_i D^{-1} x_i'\} \right. \\ &- \sum_{\alpha=1}^{N} x_i D^{-1} x_\alpha \{[x_\alpha D^{-1}(\hat{X} - X)']^2 \\ &\left. - x_\alpha D^{-1} x_\alpha'\} \frac{(1 - p_\alpha)(1 - 2p_\alpha)}{p_\alpha^2} \right\rangle \qquad 1 \le i \le N \qquad (14.20)\end{aligned}$$

In (14.20) we deleted terms of smaller order than those retained. These terms correspond to $\mu_3(\hat{X} - X)/D$ and o(1) in (14.16).

Now, setting

$$\pi_{ij} - \pi_i \pi_j = \text{Cov}(I_i, I_j | \hat{X}_h = X_h,\ 1 \le h \le m) \qquad 1 \le i \ne j \le N \qquad (14.21)$$

we must realize that the formula is senseless if there is no sample satisfying $\hat{X}_h = X_h$, $1 \le h \le m$. What is even worse, the covariances $\pi_{ij} - \pi_i \pi_j$ may be very irregular, even if the set of samples $\Omega = \{s: \hat{X}_h(s) = X_h,\ 1 \le h \le m\}$ is not empty. The only case worth considering is when the $\hat{X}_h$ are the strata sample sizes, $\hat{X}_h = K_h$ = size$(s \cap S_h)$, where the strata $\{S_h,\ 1 \le h \le m\}$ may be disjoint as well as overlapping.

CONJECTURE 14.1 Consider an arbitrary Poisson sampling and an arbitrary system of strata $\{S_1, \ldots, S_m\}$, $S_h \subset S$, $1 \le h \le m$. Put K_h = size$(s \cap S_h)$ and $n_h = EK_h$, $1 \le h \le m$. Assuming that the expectations n_h are integers, introduce $\pi_i = E(I_i | K_h = n_h,\ 1 \le h \le m)$ and $\pi_{ij} = E(I_i I_j | K_h = n_h,\ 1 \le h \le m)$.

Then

$$\pi_i = p_i\Big\{1 - \frac{1}{2}(1 - p_i)\Big[(1 - 2p_i)v_iD^{-1}v_i' - \sum_{\alpha=1}^{N}(v_iD^{-1}v_\alpha')(v_\alpha D^{-1}v_\alpha')p_\alpha(1 - p_\alpha)(1 - 2p_\alpha)\Big] + o\Big(\frac{1}{d}\Big)\Big\} \qquad 1 \le i \le N \tag{14.22}$$

where D^{-1} is the inverse (or the generalized inverse) of the matrix

$$D = \Big[\sum_{i=1}^{N} v_{hi}v_{ki}p_i(1 - p_i)\Big]_{h,k=1}^{m} \tag{14.23}$$

$$d = \text{minimal nonzero eigenvalue of } D \tag{14.24}$$

and v_i is the vector $(v_{1i},\ldots,v_{mi})$ with

$$v_{hi} = \begin{cases} 1 & \text{if } i \in S_h \\ 0 & \text{otherwise} \end{cases} \tag{14.25}$$

Furthermore,

$$\pi_i\pi_j - \pi_{ij} \sim (1 - \lambda_i)(1 - \lambda_j)\pi_i\pi_jv_iD_\lambda^{-1}v_j' \qquad 1 \le i \ne j \le N \text{ as } d \to \infty \tag{14.26}$$

where

$$D_\lambda = \Big[\sum_{i=1}^{N} v_{hi}v_{ki}\pi_i(1 - \lambda_i)\Big]_{h,k=1}^{m} \tag{14.27}$$

and the coefficients $1 - \lambda_i$ are solutions of

$$(1 - \lambda_i)[1 - \pi_i(1 - \lambda_i)v_iD_\lambda^{-1}v_i'] = 1 - \pi_i \qquad 1 \le i \le N \tag{14.28}$$

Relation (14.26) is uniform in $1 \le i < j \le N$.

Remark 14.2 The right-hand side of (14.26) is asymptotically equivalent to the residual covariance of I_i and I_j, resulting from their linear regression on $(K_1,\ldots,K_m)$. This may be seen from

(14.20) applied to $X_h = K_h$, $1 \le h \le m$. However, it has been adapted so as to be tight, i.e., to satisfy the following properties of true values:

$$0 = \mathrm{Cov}(I_i, K_h | K_h = n_h,\ 1 \le h \le m) = -\sum_{j \ne i}^{N} (\pi_i \pi_j - \pi_{ij}) v_{hj}$$
$$+ \pi_i (1 - \pi_i) v_{hi} \qquad 1 \le h \le m \qquad 1 \le i \le N \qquad (14.29)$$

Example 14.5 Consider uniform sampling with $p_i = p$, $1 \le i \le N$, and two complete sets of disjoint strata $\{S_1, \ldots, S_f\}$ and $\{T_1, \ldots, T_g\}$. We shall assume that the joint partition is uniform in the following sense:

$$\mathrm{size}(S_h \cap T_k) = \frac{N}{fg} \qquad 1 \le h \le f \qquad 1 \le k \le g \qquad (14.30)$$

Put $K_h = \mathrm{size}(s \cap S_h)$, $1 \le h \le f$, $L_k = \mathrm{size}(s \cap T_k)$, $1 \le k \le g$. Setting $K_{f+k} = L_k$ and $m = f + g$, we have a special case of the situation considered in Conjecture 14.1. Assuming that $n_h = EK_h = Np/f$, $1 \le h \le f$, and $r_k = EL_k = Np/g$, $1 \le k \le g$, are integers, let us put $\pi_i = E(I_i | K_h = n_h,\ L_k = r_k,\ 1 \le h \le f,\ 1 \le k \le g)$, $1 \le i \le N$. In view of the symmetry, all π_i terms must be the same, i.e., we have

$$\pi_i = p \qquad 1 \le i \le N \qquad (14.31)$$

even conditionally. As for the conditional covariances of I_i and I_j, $\pi_{ij} - \pi_i \pi_j$, we conjecture that they are asymptotically equivalent to residual covariance under linear regression. However, putting $K = \mathrm{size}\ s$, it is easy to see that the linear predictor for I_i equals

$$\hat{\pi}_i = p + \frac{f}{N}(K_h - n_h) + \frac{g}{N}(L_k - r_k) - \frac{1}{N}(K - Np)$$
$$\text{if } i \in S_h \cap T_k \qquad 1 \le h \le f \qquad 1 \le k \le g \quad (14.32)$$

Thus, obviously,

$$\pi_i \pi_j - \pi_{ij} \sim -\mathrm{Cov}(\hat{\pi}_i, \hat{\pi}_j) = \frac{p(1-p)}{N}\left[f \sum_{h=1}^{f} v_{hi} v_{hj} + g \sum_{k=1}^{g} u_{ki} u_{kj} - 1\right]$$
$$1 \le i \ne j \le N \qquad (14.33)$$

TABLE 14.2 Accuracy of Eq. (14.36) for N = 20, f = g = 2, p = 1/2

$\Sigma_{h=1}^{2} v_{hi}v_{hj} + \Sigma_{k=1}^{2} u_{ki}u_{kj}$	$\pi_i\pi_j - \pi_{ij}$	Approximation based on Eq. (14.36)
2	.04409	.04412
1	.01473	.01471
0	-.01473	-.01471

where

$$v_{hi} = \begin{cases} 1 & \text{if } i \in S_h \\ 0 & \text{otherwise} \end{cases} \qquad 1 \le h \le f \qquad 1 \le i \le N \quad (14.34)$$

and

$$u_{ki} = \begin{cases} 1 & \text{if } i \in T_k \\ 0 & \text{otherwise} \end{cases} \qquad 1 \le k \le g \qquad 1 \le i \le N \quad (14.35)$$

The tight version corresponding to (14.26) is

$$\pi_i\pi_j - \pi_{ij} \sim \frac{p(1-p)}{N-f-g+1}\left[f \sum_{h=1}^{f} v_{hi}v_{hj} + g \sum_{k=1}^{g} u_{ki}u_{kj} - 1\right] \qquad 1 \le i \ne j \le N \qquad (14.36)$$

The accuracy of formula (14.36) is very good as is apparent from two numerical examples in Tables 14.2 and 14.3. As we have f = g in these examples, the value of $\pi_i\pi_j - \pi_{ij}$ depends only on the number of strata that include both the units i and j. This number may be expressed as $\Sigma_{h=1}^{2} v_{hi}v_{hj} + \Sigma_{k=1}^{2} u_{ki}u_{kj}$. In Problem 14.3 there is given the formula for the exact values of $\pi_i\pi_j - \pi_{ij}$ for the case considered in Table 14.2.

TABLE 14.3 Accuracy of Eq. (14.36) for N = 24, f = g = 2, p = 1/3

$\Sigma_{h=1}^{2} v_{hi}v_{hj} + \Sigma_{k=1}^{2} u_{ki}u_{kj}$	$\pi_i\pi_j - \pi_{ij}$	Approximation based on Eq. (14.36)
2	.03167	.03175
1	.01064	.01058
0	-.01064	-.01058

PROBLEMS AND NOTES

14.1 Under Poisson sampling with parameters $p_1, \ldots, p_N$ the quadratic regression predictor for $I_i I_j$ by K = size s equals

$$\hat{\pi}_{ij} = p_i p_j \left\{ 1 + \frac{2 - p_i - p_j}{d} (K - n) + \frac{(K - n)^2 - d}{d^2} [(1 - p_i)(\bar{\bar{p}} - p_i) + (1 - p_j)(\bar{\bar{p}} - p_j) + (1 - p_i)(1 - p_j)] + o\left(\frac{1}{d}\right) \right\} \quad 1 \le i \ne j \le N \quad (14.37)$$

where $n = EK$, $\bar{\bar{p}} = \Sigma_{i=1}^{N} p_i^2(1 - p_i)/d$, $d = \Sigma_{i=1}^{N} p_i(1 - p_i)$, and $d\, o(1/d) \to 0$ in probability.

14.2 If $S_1 \cap S_2 = \emptyset$, then $K_1 = \text{size}(s \cap S_1)$ and $K_2 = \text{size}(s \cap S_2)$ are independent random variables under Poisson sampling.

14.3 Consider the situation of Example 14.5 with N = 4r, n = 2r, and f = g = 2. Then the total number of samples s satisfying the conditions $K_1 = K_2 = L_1 = L_2 = r$ equals

$$M = \sum_{\alpha=0}^{r} \binom{r}{\alpha}^4$$

If we denote the number of strata including both the units i and j by c, then the number of samples satisfying $K_1 = K_2 = L_1 = L_2 = r$ and moreover including the units i and j equals

$$M_{ij} = \begin{cases} \sum_{\alpha=2}^{r} \binom{r-2}{\alpha-2}\binom{r}{\alpha}^3 & \text{if } c = 2 \\ \sum_{\alpha=1}^{r-1} \binom{r-1}{\alpha-1}\binom{r-1}{\alpha}\binom{r}{\alpha}^2 & \text{if } c = 1 \\ \sum_{\alpha=1}^{r} \binom{r-1}{\alpha-1}^2\binom{r}{\alpha}^2 & \text{if } c = 0 \end{cases}$$

We have $\pi_{ij} = M_{ij}/M$.

14.4 The formula (14.13) will be "tight" if

$$(1 - \lambda_i)\left[1 - \frac{\pi_i(1 - \lambda_i)}{d_\lambda} + 2\,\frac{\pi_i(1 - \lambda_i)(\pi_i - \bar{\bar{\pi}})^2}{d_\lambda^2}\right] = 1 - \pi_i$$

$$1 \le i \le N$$

where $d_\lambda = \Sigma_{i=1}^N \pi_i(1 - \lambda_i)$.

15

CORRECTING A SAMPLE

Assume that we have a sample s' obtained by a sampling method with fixed size n' and inclusion probabilities π'_i, $1 \le i \le N$, and that we want to transform it into a sample s, which might be considered as obtained by a different sampling method, with fixed size n and inclusion probabilities π_i, $1 \le i \le N$. Our aim is to take s as little different from s' as possible. According to (3.10), we have

$$\sum_{i=1}^{N} \pi'_i = n' \qquad \sum_{i=1}^{N} \pi_i = n \tag{15.1}$$

Case 1. Assume that $\pi'_i \ge \pi_i$, $1 \le i \le N$. Then our task will be approximately fulfilled by removing from s' a rejective sample of size $n_\Delta = n' - n$ and with inclusion probabilities π_i^Δ proportionate to $(\pi'_i - \pi_i)/\pi'_i$, i.e.,

$$\pi_i^\Delta = \frac{(n' - n)(\pi'_i - \pi_i)}{\pi'_i}\left(\sum_{j\in s'} \frac{\pi'_j - \pi_j}{\pi'_j}\right)^{-1} \qquad i \in s' \tag{15.2}$$

Denoting the just-described rejective sample by s_Δ and putting

$$s = s' - s_\Delta \tag{15.3}$$

we have

$$\pi_i^* = P(i \in s) = P(i \in s')P(i \notin s_\Delta | i \in s') \tag{15.4}$$

Now

$$P(i \notin s_\Delta | i \in s') = E(1 - \pi_i^\Delta | i \in s')$$

$$= 1 - \frac{(n' - n)(\pi_i' - \pi_i)}{\pi_i'} E\left[\left(\sum_{j \in s'} \frac{\pi_j' - \pi_j}{\pi_j'}\right)^{-1} \middle| i \in s'\right] \tag{15.5}$$

If the size of s' is large enough, the presence or absence of the unit i in s' will have little effect on the sum $\Sigma_{j \in s'}(\pi_j' - \pi_j)/\pi_j'$. Thus we are allowed to assume that

$$E\left[\left(\sum_{j \in s'} \frac{\pi_j' - \pi_j}{\pi_j'}\right)^{-1} \middle| i \in s'\right] \sim E\left(\sum_{j \in s'} \frac{\pi_j' - \pi_j}{\pi_j'}\right)^{-1} \tag{15.6}$$

Furthermore, if n' is large enough, the relative variability of the same sum will be small and we may write

$$E\left(\sum_{j \in s'} \frac{\pi_j' - \pi_j}{\pi_j'}\right)^{-1} \sim \left(E \sum_{j \in s'} \frac{\pi_j' - \pi_j}{\pi_j'}\right)^{-1}$$

$$= \left(\sum_{j=1}^{N} \frac{\pi_j' - \pi_j}{\pi_j'} \pi_j'\right)^{-1} = (n' - n)^{-1} \tag{15.7}$$

Consequently, returning to (15.4), we have

$$\pi_i^* \sim \pi_i' \left[1 - \frac{(n' - n)(\pi_i' - \pi_i)}{\pi_i'} (n' - n)^{-1}\right] = \pi_i \tag{15.8}$$

Q.E.D.

Case 2. Assume that $\pi_i' \le \pi_i$, $1 \le i \le N$. Then our task will be approximately fulfilled by drawing from $r' = S - s'$ a rejective sample of size $n_\Delta = n - n'$ with inclusion probabilities π_i^Δ proportionate to $(\pi_i - \pi_i')/(1 - \pi_i')$, i.e.,

$$\pi_i^\Delta = \frac{(n - n')(\pi_i - \pi_i')}{1 - \pi_i'} \left(\sum_{j \in r'} \frac{\pi_j - \pi_j'}{1 - \pi_j'}\right)^{-1} \qquad i \in r' \tag{15.9}$$

Denoting the last-named rejective sample again by s_Δ and putting

$$s = s' \cup s_\Delta \tag{15.10}$$

we shall have

$$\pi_i^* = P(i \in s) = P(i \in s') + P(i \in r')P(i \in s_\Delta | i \in r')$$

$$= \pi_i' + (1 - \pi_i')\frac{(n - n')(\pi_i - \pi_i')}{1 - \pi_i'} E\left[\left(\sum_{j \in r'} \frac{\pi_j - \pi_j'}{1 - \pi_j'}\right)^{-1} \middle| i \in r'\right]$$

$$= \pi_i' + (n - n')(\pi_i - \pi_i')E\left[\left(\sum_{j \in r'} \frac{\pi_j - \pi_j'}{1 - \pi_j'}\right)^{-1} \middle| i \in r'\right] \tag{15.11}$$

Now, again for large r', we may write

$$E\left[\left(\sum_{j \in r'} \frac{\pi_j - \pi_j'}{1 - \pi_j'}\right)^{-1} \middle| i \in r'\right] \sim \left(\sum_{j=1}^{N} \frac{\pi_j - \pi_j'}{1 - \pi_j'}(1 - \pi_j')\right)^{-1}$$

$$= (n - n')^{-1}$$

and

$$\pi_i^* \sim \pi_i' + (n - n')(\pi_i - \pi_i')(n - n')^{-1} = \pi_i \tag{15.12}$$

Q.E.D.

Case 3. If neither case 1 nor case 2 holds, then we divide the population into three strata:

$$\begin{aligned} S_+ &= \{i: \pi_i' > \pi_i\} \\ S_0 &= \{i: \pi_i' = \pi_i\} \\ S_- &= \{i: \pi_i' < \pi_i\} \end{aligned} \tag{15.13}$$

Furthermore, we divide the sample s' into three disjoint subsamples:

$$s_+' = s' \cap S_+ \qquad s_0' = s' \cap S_0 \qquad s_-' = s' \cap S_- \tag{15.14}$$

Next we put $r_-' = S - s_-'$. Then we draw from s_+' a rejective sample s_Δ^+ of size

$$n_\Delta^+ = \sum_{i \in S_+} (\pi_i' - \pi_i) \tag{15.15}$$

and with inclusion probabilities

$$\pi_{i+}^{\Delta} = \frac{n_{\Delta}^{+}(\pi_i' - \pi_i)}{\pi_i'}\left(\sum_{j\in s_+'} \frac{\pi_j' - \pi_j}{\pi_j'}\right)^{-1} \qquad i \in s_+' \tag{15.16}$$

and from r_-' a sample s_{Δ}^{-} of size

$$n_{\Delta}^{-} = \sum_{i\in S_-} (\pi_i - \pi_i')$$

and with inclusion probabilities

$$\pi_{i-}^{\Delta} = \frac{n_{\Delta}^{-}(\pi_i - \pi_i')}{1 - \pi_i'}\left(\sum_{i\in r_-'} \frac{\pi_i - \pi_i'}{1 - \pi_i'}\right)^{-1} \qquad i \in r_-' \tag{15.17}$$

Finally, we put

$$s = (s' \cup s_{\Delta}^{-}) - s_{\Delta}^{+} \tag{15.18}$$

The approximate relation $\pi_i^* = P(i \in s) \sim \pi_i$ is then obtained by applying case 1 and case 2 to the strata S_+ and S_-, respectively.

Remark 15.1 Even if $\pi_i^* \sim \pi_i$ holds only approximately, the relation

$$\sum_{i=1}^{N} \pi_i^* = \sum_{i=1}^{N} \pi_i \tag{15.19}$$

holds exactly, in view of (15.1) and the fact that the π_i^* terms refer to sampling of fixed size n. Thus the two approximations used in (15.6) and (15.7) compensated each other as far as totals are concerned.

Remark 15.2 If n_{Δ}^{+} or n_{Δ}^{-} are not integers, we may round them off to the nearest integer, keeping in mind that the rounding process must not change the difference $n_{\Delta}^{-} - n_{\Delta}^{+}$.

Remark 15.3 One might ask why we should apply just rejective sampling for selecting the differential samples $s_{\Delta}(s_{\Delta}^{+}, s_{\Delta}^{-})$. The only reason is our belief that rejective sampling is the most natural extension of simple random sampling to variable π_i terms. Besides, for rejective sampling we are able to support the procedure of Example 15.1 below by a limit theorem (see Hájek, 1964). Of course, practically, rejective sampling may be replaced by successive sampling with even success.

Example 15.1 Consider Poisson sampling with expected size n and a Poisson sample s' of actual size n' ≠ n. Then conditioning on size s' = n', we have

$$\pi_i' = P(i \in s' | \text{size } s' = n')$$

$$\sim p_i \left[1 + \frac{(n' - n)}{d}(1 - p_i) + 0\left(\frac{1}{d}\right)\right] \tag{15.20}$$

where $d = \Sigma_{i=1}^N p_i(1 - p_i)$. Note that formula (15.20) follows from (7.36) when we drop terms of order $0(1/d)$. Now if we want to transform s' into a sample corresponding to $\pi_i = p_i$, $1 \le i \le N$, we have case 1 for $n' > n$ and case 2 for $n' < n$. Thus, by (15.2),

$$\pi_i^\Delta \doteq \frac{(n' - n)(1 - p_i)}{\Sigma_{j \in s'} (1 - p_j)} \qquad \text{if } n' > n \tag{15.21}$$

and, by (15.9),

$$\pi_i^\Delta \doteq \frac{(n - n')p_i}{\Sigma_{j \in r'} p_j} \qquad \text{if } n' < n \tag{15.22}$$

Note that under the circumstances just described case 1 and case 2 correspond to the method of obtaining a rejective sample of size n from a Poisson sample of expected size n as described in Example 7.5.

Example 15.2 Let s' be a Poisson sample with parameters p_1, ..., p_N, $\Sigma_{i=1}^N p_i = n$, conditioned by size s' = n. Then s' may be regarded as a rejective sample with inclusion probabilities π_i' where, according to Theorem 7.3,

$$\pi_i' = p_i\left[1 - \frac{(\bar{\bar{p}} - p_i)(1 - p_i)}{d} + o\left(\frac{1}{d}\right)\right] \qquad 1 \le i \le N \tag{15.23}$$

If we want to transform s' into a rejective sample s corresponding to the inclusion probabilities $\pi_i = p_i$, $1 \le i \le N$, we have case 3. However,

$$n_\Delta^+ + n_\Delta^- = \sum_{i=1}^{N} |\pi_i' - \pi_i| \doteq \sum_{i=1}^{N} |\bar{\bar{p}} - p_i| \frac{p_i(1 - p_i)}{d}$$

$$\le \max_{1\le i\le N} p_i \tag{15.24}$$

Thus, if, for example, $\max_{1\le i\le N} p_i < 0.2$, we shall have $n_\Delta^+ < 0.1$, and approximating by integers, we shall put $n_\Delta^+ = n_\Delta^- = 0$.

Consequently we shall carry out no correction at all. In other words, rejective sampling with $\pi_i = nx_i/X$ may usually be approximated by Poisson sampling with $p_i = nx_i/X$ conditioned by size s = n. Note that formula (15.23) differs from (15.20) only by the explicit expression of terms of order $O(1/d)$.

Example 15.3 If n/N is not very large, then the inclusion probabilities corresponding to drawing probabilities $\alpha_i = x_i/X$ equal approximately

$$\pi_i' = \frac{nx_i}{X}\left[1 - (n - 1)\frac{x_i}{X} + (n - 1)\sum_{j=1}^{N}\frac{x_j^2}{X^2}\right] \tag{15.25}$$

for rejective sampling, and

$$\pi_i' = \frac{nx_i}{X}\left[1 - \frac{1}{2}(n - 1)\frac{x_i}{X} + \frac{1}{2}(n - 1)\sum_{j=1}^{N}\frac{x_j^2}{X^2}\right] \tag{15.26}$$

for successive sampling. If we wanted to have $\pi_i = nx_i/X$, $1 \le i \le N$, the total size of the correction equals

$$\sum_{i=1}^{N} |\pi_i - \pi_i'| = \sum_{i=1}^{N} \pi_i \left|\pi_i - \frac{1}{n}\sum_{j=1}^{N} \pi_j^2\right|$$

for rejective sampling, and must be halved for successive sampling. Thus, if we hesitate whether we can put $\alpha_i = x_i/X$, it suffices to estimate $\Sigma_{i=1}^N |\pi_i - \pi_i'|$ according to the above formula. If it is small, say $\Sigma|\pi_i - \pi_i'| < 1$, we can ignore the necessity to revise the α values. For data of Example 9.4, where we have a successive sample of size 50, one can compute

$$\sum_{i=1}^{1000} |\pi_i - \pi_i'| = 1.03$$

Dividing by 2, we obtain a number smaller than 1, which suggests that we could drop revising the α_i terms. The actual course of sampling confirms this conclusion.

PROBLEMS AND NOTES

15.1 Assume that we have a sample of size 100 from the population of Table 6.1 with π_i terms proportionate to the x_i terms, and that we wish to transform it into a sample of the same size but with π_i terms proportionate to $x_i^{0.6}$. Establish n_Δ^+ and n_Δ^- for this operation. *Hint:* Utilize data for f_k in Tables 6.3 and 6.6.

15.2 Elaborate formulas for transforming a Poisson sample into a stratified rejective sample.

Part III

METHODS OF ESTIMATION

16

REPRESENTATIVENESS, ROBUSTNESS, AND TIGHTNESS

As we know from Chapter 4 the general linear estimate $\hat{Y}$ for the population total Y has the form

$$\hat{Y} = \sum_{i \in s} y_i w_i(s) = \sum_{i=1}^{N} y_i w_i(s) \tag{16.1}$$

where the weights $w_i(s)$ vanish for $i \notin s$. In what follows we shall abbreviate $w_i(s)$ as w_i. A sampling-estimating strategy, say (P,w), is defined by probabilities $\{P(s),\ s \subset S\}$ and weights $\{w_i(s),\ s \subset S,\ i \in s\}$. Since $\hat{Y} - Y = \Sigma_{i=1}^{N} y_i(w_i - 1)$, the mean square error equals

$$E(\hat{Y} - Y)^2 = \sum_{i=1}^{N} y_i^2 E(w_i - 1)^2 + \sum_{i \neq j}^{N} \sum^{N} y_i y_j c_{ij} \tag{16.2}$$

where

$$c_{ij} = E[(w_i - 1)(w_j - 1)] \tag{16.3}$$

A rational choice of the strategy (P,w) will depend, in the first place, on our prior knowledge of one or several auxiliary variables that are correlated with the variable y under issue. Denoting these variables by x_k, $1 \leq k \leq m$, and their values by x_{ki}, $1 \leq i \leq N$, the presence of correlation means that the residuals

$$e_i = y_i - \sum_{k=1}^{m} \beta_k x_{ki} \qquad 1 \leq i \leq N \tag{16.4}$$

are significantly less variable than the values y_i themselves, provided the regression parameters β_k are properly chosen. Variables x_k will be called regressors (predictor variables, explanatory variables), and the model (16.4) will be called the regression model. The population totals of the variables x_k will be denoted by X_k,

$$X_k = \sum_{i=1}^{N} x_{ki} \qquad 1 \le k \le m \tag{16.5}$$

Given some weights t_i, $1 \le i \le N$, the variability of e_i may be measured by

$$\sum_{i=1}^{N} e_i^2 t_i = \sum_{i=1}^{N} \left(y_i - \sum_{k=1}^{m} \beta_k x_{ki}\right)^2 t_i \tag{16.6}$$

and proper values of β_k, say $\bar{\beta}_k$, may be those minimizing the sum of squares. We shall call $\bar{\beta}_k$ the *least-squares population estimates of* β_k *with weights* t_i and put

$$\bar{e}_i = y_i - \sum_{k=1}^{m} \bar{\beta}_k x_{ki} \qquad 1 \le i \le N \tag{16.7}$$

The symbol β_k will be reserved for unknown parameters utilized by "nature," when generating the values y_i by means of regressors x_{ki} and random fluctuations.

Example 16.1 The knowledge of the identity of every variable may be characterized by the regressors $x_i = 1$, $1 \le i \le N$. The residuals equal

$$e_i = y_i - \beta \tag{16.8}$$

and the least-squares population estimate $\bar{\beta}$ of β with weights $t_i = 1$ equals the population average:

$$\bar{\beta} = \bar{Y} \tag{16.9}$$

Hence

$$\bar{e}_i = y_i - \bar{Y} \qquad 1 \le i \le N \tag{16.10}$$

The total X equals N, the size of the population.

Example 16.2 Consider a partition of S into L strata S_h, $1 \le h \le L$, and introduce regression model (2.5) in each stratum separately. Then we have

$$e_i = y_i - \beta_h \qquad i \in S_h \qquad 1 \le h \le L \tag{16.11}$$

The least-squares stratum estimate with weights $t_i = 1$ for β will obviously equal the stratum average $\bar{Y}_h$. Thus,

$$\bar{e}_i = y_i - \bar{Y}_h \qquad i \in S_h \qquad 1 \le h \le L \tag{16.12}$$

The model may be formulated within the whole population by means of indicator variables u_h for individual strata:

$$u_{hi} = \begin{cases} 1 & \text{if } i \in S_h \\ 0 & \text{otherwise} \end{cases} \tag{16.13}$$

Then (16.11) may be rewritten as follows:

$$e_i = y_i - \sum_{h=1}^{L} \beta_h u_{hi} \qquad 1 \le i \le N \tag{16.14}$$

Consequently, within the whole population, the regression model involves L regression parameters.

Example 16.3 Let x_i be some measure of size of the unit i and let the y_i terms be proportionate, up to random fluctuations, to the size. For example, x_i may denote the number of subunits included in the unit i, and y_i may be the total of values ascertained on these subunits. If the unit is a farm, x_i may be the area under wheat and y_i the total yield of wheat; or if the unit is a town, x_i may be the number of inhabitants in some past year and y_i the number of inhabitants of some special kind (say male agricultural workers) in the present year. Then proper residuals equal

$$e_i = y_i - \beta x_i \qquad 1 \le i \le N \tag{16.15}$$

(the case of linear regression passing through the origin).

The least-squares population estimate of β with weights $t_i = 1/x_i$ can be easily found as

$$\bar{\beta} = \frac{Y}{X} \tag{16.16}$$

Hence,

$$\bar{e}_i = y_i - \frac{Y}{X} x_i \qquad 1 \le i \le N \tag{16.17}$$

Example 16.4 Let x_i be the measurements of the same variable as y_i, but carried out on a past occasion or by a less precise method. Then (16.17) is again applicable, and $\bar{\beta} = Y/X$ will be close to 1.

If we feel that the residuals $e_i = y_i - \Sigma_{k=1}^{m} \beta_k x_{ki}$ are significantly less variable than the original values y_i, we should require that the sampling error $\hat{Y} - Y$ depend only on e_i, $1 \le i \le N$. Obviously, we shall have, for any parameters β_k,

$$\hat{Y} - Y = \sum_{i=1}^{N} y_i(w_i - 1) = \sum_{i=1}^{N} \left\{y_i - \sum_{k=1}^{m} \beta_k x_{ki}\right\}(w_i - 1)$$
$$= \sum_{i=1}^{N} e_i(w_i - 1) \tag{16.18}$$

if and only if

$$\sum_{i\in s} x_{ki} w_i = \sum_{i=1}^{N} x_{ki} \qquad 1 \le k \le m \tag{16.19}$$

holds for any sample s that may be selected under the sampling design P. Relations (16.19) mean that $\hat{Y} = Y$ holds with probability 1 as soon as the y_i terms are exactly a linear combination of x_{ki}, $1 \le k \le m$, i.e., if

$$y_i = \sum_{k=1}^{m} \beta_k x_{ki} \qquad 1 \le i \le N \tag{16.20}$$

holds for some parameters β_k.

DEFINITION A strategy (P,w) satisfying (16.19) will be called *representative* with respect to the regressors x_k.

Equations (16.19) provide the mathematical essence of the term *representative sampling* in circulation some decades ago.

Representativeness has important consequences for the constants c_{ij} defined by (16.3). Rewriting (16.19) as $\hat{X}_k = X_k$, $1 \le k \le m$, we obtain that the mean square error vanishes as soon as $y_i = x_{ki}$:

$$0 = E(\hat{X}_k - X_k)^2 = \sum_{i=1}^{N} x_{ki}^2 E(w_i - 1)^2 + \sum_{i \neq j}^{N} \sum_{j=1}^{N} x_{ki} x_{kj} c_{ij} \qquad 1 \le k \le m \tag{16.21}$$

Moreover, since also $E(w_i - 1)(\hat{X}_k - X_k) = 0$ holds for all i, $1 \le i \le N$, we have

$$E(w_i - 1)(\hat{X}_k - X_k) = \sum_{j=1}^{N} x_{kj} E(w_i - 1)(w_j - 1) = 0$$

that is,

$$\sum_{j \neq i}^{N} x_{kj} c_{ij} + x_{ki} E(w_i - 1)^2 = 0 \qquad 1 \le i \le N,\ 1 \le k \le m \tag{16.22}$$

For any strategy (P,w) the mean square error is a quadratic form

$$y'Cy = \sum_{i=1}^{N} \sum_{j=1}^{N} y_i y_j c_{ij} \tag{16.23}$$

where $c_{ii} = E(w_i - 1)^2$. For representative strategies this form vanishes, $y'Cy = 0$, as soon as (16.20) holds.

DEFINITION An approximation for $E(\hat{Y} - Y)^2$, say $\bar{E}(\hat{Y} - Y)^2$, will be called *tight* if it shares with $E(\hat{Y} - Y)^2$ both the above properties. Thus a quadratic form

$$\bar{E}(\hat{Y} - Y)^2 = \sum_{i=1}^{N} \sum_{j=1}^{N} y_i y_j \bar{c}_{ij}$$

will be a tight approximation for (16.23) if

$$\bar{c}_{ii} = E(w_i - 1)^2 \qquad 1 \le i \le N \tag{16.24}$$

and

$$\bar{E}(\hat{X}_k - X_k)^2 = \sum_{i=1}^{N} \sum_{j=1}^{N} x_{ki} x_{kj} \bar{c}_{ij} = 0 \qquad 1 \le k \le m \tag{16.25}$$

Property (16.25) is equivalent to

$$\sum_{j \ne i}^{N} x_{kj} \bar{c}_{ij} + x_{ki} E(w_i - 1)^2 = 0 \qquad 1 \le i \le N,\ 1 \le k \le m \tag{16.26}$$

which is analogous to (16.22).

The following theorem describes a family of tight approximate formulas for the mean square error. This family will find numerous applications for sampling-estimating strategies considered in the sequel.

THEOREM 16.1 Let $\bar{\beta}_k$ be the values of β_k that minimize (16.6) for some positive weights t_i, $1 \le k \le m$, $1 \le i \le N$. Then the formula

$$\bar{E}(\hat{Y} - Y)^2 = \sum_{i=1}^{N} \left(y_i - \sum_{k=1}^{m} \bar{\beta}_k x_{ki} \right)^2 v_i \tag{16.27}$$

is tight if

$$v_i = E(w_i - 1)^2 \left[1 - 2 \sum_{k=1}^{m} a_{ki} x_{ki} + \sum_{k=1}^{m} \sum_{h=1}^{m} a_{ki} a_{hi} \sum_{\alpha=1}^{N} \frac{x_{k\alpha} x_{h\alpha} v_\alpha}{v_i} \right]^{-1} \qquad 1 \le i \le N \tag{16.28}$$

where the constants a_{ki} are defined by

$$\bar{\beta}_k = \sum_{i=1}^{N} a_{ki} y_i \tag{16.29}$$

Proof. If $y_i = x_{ki}$, the minimum value of (16.6) is zero and is obtained for $\bar{\beta}_h = 1$ and $\bar{\beta}_k = 0$, $k \ne h$. Then $y_i - \Sigma_{k=1}^{m} \bar{\beta}_k x_{ki} =$

$x_{hi} - x_{hi} = 0$ holds for every $i = 1, \ldots, N$, and consequently $\bar{E}(\hat{X}_h - X_h)^2$, $1 \le h \le m$. Thus (16.25) is satisfied for every weight v_i.

Next, substituting (16.29) into (16.27), we easily derive that under (16.28) the coefficient standing at y_i^2 equals $E(w_i - 1)^2$. Consequently, (16.24) is satisfied. Q.E.D.

If m = 1, (16.27) becomes

$$\bar{E}(\hat{Y} - Y)^2 = \sum_{i=1}^{N} (y_i - \bar{\beta}x_i)^2 v_i \tag{16.30}$$

where

$$\bar{\beta} = \frac{\sum_{i=1}^{N} y_i x_i t_i}{\sum_{i=1}^{N} x_i^2 t_i} \tag{16.31}$$

and

$$v_i = E(w_i - 1)^2 \left[1 - 2\frac{x_i^2 t_i}{\sum_{\alpha=1}^{N} x_\alpha^2 t_\alpha} + \frac{x_i^2 t_i^2 / v_i}{(\sum_{\alpha=1}^{N} x_\alpha^2 t_\alpha)^2} \sum_{\alpha=1}^{N} x_\alpha^2 v_\alpha\right]^{-1} \tag{16.32}$$

In the special case $t_i = 1/x_i$, we have

$$v_i = E(w_i - 1)^2 \left(1 - 2\frac{x_i}{X} + \frac{\sum_{\alpha=1}^{N} x_\alpha^2 v_\alpha / v_i}{X^2}\right)^{-1} \qquad 1 \le i \le N \tag{16.33}$$

and if $t_i = \lambda v_i$, then

$$v_i = E(w_i - 1)^2 \left(1 - \frac{x_i^2 v_i}{\sum_{\alpha=1}^{N} x_\alpha^2 v_\alpha}\right)^{-1} \qquad 1 \le i \le N \tag{16.34}$$

As we can see, v_i equals approximately $E(w_i - 1)^2$ up to a multiplying factor, which differs from 1 only very little if N is large. In (16.34) the harmonic mean of the factor equals $N/(N - 1)$, since

$$\left[\frac{1}{N}\sum_{i=1}^{N}\left(1 - \frac{x_i^2 v_i}{\sum_{\alpha=1}^{N} x_\alpha^2 v_\alpha}\right)\right]^{-1} = \left(\frac{N - 1}{N}\right)^{-1} = \frac{N}{N - 1}$$

If $t_i \neq \lambda v_i$, the harmonic mean of the multiplying factor will tend to be somewhat larger: it will equal $N/(N - 1 + V_*^2)$, where V_* is the weighted correlation coefficient of t_i/v_i with weights $x_i v_i$, $1 \leq i \leq N$. This follows from (16.32) if we apply (1.21).

In the case of m regressors and $t_i = \lambda v_i$ the harmonic mean of the multiplying factor can be shown to be $N/(N - m)$; if $t_i \neq \lambda v_i$, it will be somewhat larger.

Strategies (P,w) for which formula (16.27) is applicable will be called *robust*. For robust strategies the mean square error will be well estimable whatever the population $(y_1,\ldots,y_N)$ may be and the precision of error estimates will be comparable to the precision achieved in models based on independent observations. The source of this phenomenon lies in the fact that in (16.27), $\bar{E}(\hat{Y} - Y)^2$ depends only on the squares of the residuals $\bar{e}_i^2$, whereas the error-products $\bar{e}_i\bar{e}_j$ are missing.

Primary differences between robust strategies will consist in the set of regressors x_k to which they are representative. Two strategies with coinciding regressors may still differ in the values of $E(w_i - 1)^2$ and in the weight t_i used in the definition of $\bar{\beta}_k$. The latter differences, however, are of secondary importance. Nonetheless, their study will be important for finer discrimination between the strategies.

The weights v_i given implicitly by the system of equations (16.28) are difficult to establish precisely. However, fortunately, there is no need to compute them. For theoretical considerations it suffices to know that they equal $E(w_i - 1)^2$ times a multiplier whose harmonic average value is $N/(N - m)$ if $t_i = \lambda v_i$ and is somewhat larger if $t_i \neq \lambda v_i$.

In estimation problems, we shall see later that in robust strategies $E(\hat{Y} - Y)^2$ may be estimated without the explicit knowledge of the v_i terms.

PROBLEMS AND NOTES

16.1 If the strategy (P,w) is representative with respect to the regressors $x_1, \ldots, x_m$, we have

$$E(\hat{Y} - Y)^2 = \sum_{i=1}^{N} \left(y_i - \sum_{k=1}^{m} \beta_k x_{ki}\right)^2 E(w_i - 1)^2$$

$$- \sum_{i \neq j}^{N} \sum^{N} \left(y_i - \sum_{k=1}^{m} \beta_k x_{ki}\right)\left(y_j - \sum_{k=1}^{m} \beta_k x_{kj}\right) c_{ij}$$

for all values $\beta_1, \ldots, \beta_m$.

16.2 The estimate

$$\hat{Y} = y + \frac{X - x}{n} \sum_{i \in s} \frac{y_i}{x_i}$$

where y and x are the sample totals, is representative relative to x under any design of fixed size n.

16.3 A linear estimate with weights w_i is unbiased for all $(y_1, \ldots, y_N)$ if and only if

$$Ew_i = 1 \qquad 1 \leq i \leq N$$

16.4 If $y'\bar{C}y$ is a tight approximation for the mean square error $y'Cy$, then we have

$$\sup \frac{y'Cy}{y'\bar{C}y} = \lambda = \frac{z'Cz}{z'\bar{C}z}$$

where λ is the maximum number for which there is a vector z such that

$$Cz = \lambda\bar{C}z$$

and z is the vector. The supremum is taken over all vectors y that are not a linear combination of regressor vectors $x_1, \ldots, x_m$.

17

SIMPLE LINEAR ESTIMATE

The simple linear estimate plays a central role in the robust theory of sampling. Ideally, if representativeness could always be easily obtained by proper sampling only, it should be the only estimate to be used. Moreover, even if we make use of more complex estimates, simple estimates have an important place as their elements. These statements will be theoretically established in Chapters 18 and 19.

Given a sampling method with positive probabilities of inclusion π_i, the simple linear estimate is defined as follows:

$$\hat{Y} = \sum_{i \in s} \frac{y_i}{\pi_i} \tag{17.1}$$

If the π_i terms are proportionate to some numbers x_i and sampling size is n, then

$$\pi_i = \frac{nx_i}{X}$$

and

$$\hat{Y} = \frac{X}{n} \sum_{i \in s} \frac{y_i}{x_i} \tag{17.2}$$

Since we shall consider linear estimates only, we shall call $\hat{Y}$ a simple estimate for short.

THEOREM 17.1 The simple estimate is unbiased provided that $\pi_i > 0$, $1 \leq i \leq N$. Moreover, if the sampling is of fixed size, $\hat{Y}$ is representative with respect to any numbers that are proportionate to the π_i terms.

Proof. Introducing the indicators of inclusion I_i by (3.2), we may write

$$\hat{Y} = \sum_{i=1}^{N} y_i \frac{I_i}{\pi_i} \tag{17.3}$$

and (3.5) entails

$$E\hat{Y} = \sum_{i=1}^{N} y_i \frac{EI_i}{\pi_i} = \sum_{i=1}^{N} y_i \frac{\pi_i}{\pi_i} = \sum_{i=1}^{N} y_i = Y$$

This proves unbiasedness.

If the sampling is of fixed size n and $x_i = \lambda\pi_i$, then, in view of (3.11),

$$\hat{X} = \sum_{i\in s} \frac{x_i}{\pi_i} = \sum_{i\in s} \lambda \frac{\pi_i}{\pi_i} = \lambda n$$

$$= \lambda \sum_{i=1}^{N} \pi_i = \sum_{i=1}^{N} x_i = X$$

holds for every sample of size n, i.e., $\hat{X} = X$ with probability 1. This proves representativeness. Q.E.D.

In particular, (17.1) is representative with respect to x under successive, rejective, or Sampford-Durbin sampling of size n.

Now we shall derive formulas for the variance of the simple estimate. Representing the simple estimate in the form $\hat{Y} = \Sigma_{i=1}^{N} y_i w_i$, (17.3) entails that

$$w_i = \frac{I_i}{\pi_i} \qquad 1 \le i \le N \tag{17.4}$$

Consequently, since $I_i^2 = I_i$,

$$E(w_i - 1)^2 = E\left(\frac{I_i}{\pi_i} - 1\right)^2 = E\left(\frac{I_i^2}{\pi_i^2} - 2\frac{I_i}{\pi_i} + 1\right) = \frac{EI_i^2}{\pi_i^2} - 2\frac{EI_i}{\pi_i} + 1$$

$$= \frac{\pi_i}{\pi_i^2} - 2\frac{\pi_i}{\pi_i} + 1 = \frac{1}{\pi_i} - 1 \qquad 1 \le i \le N \tag{17.5}$$

Furthermore,

$$c_{ij} = E[(w_i - 1)(w_j - 1)] = E\left(\frac{I_i I_j}{\pi_i \pi_j} - \frac{I_i}{\pi_i} - \frac{I_j}{\pi_j} + 1\right)$$

$$= \frac{\pi_{ij}}{\pi_i \pi_j} - 1 \qquad 1 \le i \ne j \le N \tag{17.6}$$

Substituting these values into (16.2), we obtain

$$E(\hat{Y} - Y)^2 = \sum_{i=1}^{N} y_i^2 \left(\frac{1}{\pi_i} - 1\right) + \sum_{i \ne j}^{N} \sum^{N} y_i y_j \left(\frac{\pi_{ij}}{\pi_i \pi_j} - 1\right) \tag{17.7}$$

THEOREM 17.2 Formula (17.7) may be rewritten as follows:

$$E(\hat{Y} - Y)^2 = \frac{1}{2} \sum_{i \ne j}^{N} \sum^{N} \left(\frac{y_i}{\pi_i} - \frac{y_j}{\pi_j}\right)^2 (\pi_i \pi_j - \pi_{ij}) \tag{17.8}$$

provided that sampling is of fixed size.

Proof. It suffices to compare the coefficients of y_i^2 and $y_i y_j$ in Eqs. (17.7) and (17.8). In (17.8) the coefficient of y_i^2 equals, in view of (3.12),

$$\frac{1}{\pi_i^2} \sum_{j \ne i}^{N} (\pi_i \pi_j - \pi_{ij}) = \frac{1}{\pi_i} - 1 \qquad 1 \le i \le N$$

which is concordant with the same coefficient in (17.7). The coefficient of $y_i y_j$ is $\pi_{ij}/\pi_i \pi_j - 1$, as may be easily seen. Q.E.D.

Formula (17.8) is called the Yates-Grundy formula [see Yates and Grundy (1953)].

If the probabilities of inclusion are uniform, i.e.,

$$\pi_i = \frac{n}{N} \qquad 1 \le i \le N \tag{17.9}$$

the simple estimate equals

$$\hat{Y} = \frac{N}{n} \sum_{i \in s} y_i = N\bar{y} \tag{17.10}$$

where $\bar{y}$ is the sample average. This form has the simple estimate in connection with simple random sampling of Chapter 5.

THEOREM 17.3 Under simple random sampling of size n the variance of the estimate (17.10) equals

$$E(\hat{Y} - Y)^2 = \frac{N}{n} \sum_{i=1}^{N} (y_i - \bar{Y})^2 \frac{N - n}{N - 1} \tag{17.11}$$

Proof. Substituting (5.6) into (17.8), we have

$$E(\hat{Y} - Y)^2 = \frac{1}{2} \sum_{i \neq j}^{N} \sum^{N} \left(\frac{y_i}{n/N} - \frac{y_j}{n/N} \right)^2 \frac{n(N - n)}{N^2(N - 1)}$$

$$= \frac{1}{2} \frac{1}{n} \sum_{i \neq j}^{N} \sum^{N} (y_i - y_j)^2 \frac{N - n}{N - 1} \tag{17.12}$$

Now it suffices to apply (1.32). Q.E.D.

When we sample with variable probabilities and use the simple estimate, a robust strategy is provided by any of the following three samplings: successive, rejective, or Sampford-Durbin. As we know from Chapter 16, robustness means that the exact variance formula is well approximable by (16.30) if t_i and v_i are properly chosen. We are forced to use this approximation since the exact formula is too complicated and gives no insight so far as the comparison of the methods is concerned.

In order to fit the approximate formula (16.30) to the exact formula, it is expedient to have both of them in the Yates-Grundy form. Assuming $\pi_i = nx_i/X$, we shall define the Yates-Grundy form of (16.30) as the last of the following obvious identities:

$$\sum_{i=1}^{N} (y_i - \bar{\beta}x_i)^2 v_i = \sum_{i=1}^{N} \left(y_i - \frac{\sum_{k=1}^{N} y_k \pi_k t_k}{\sum_{k=1}^{N} \pi_k^2 t_k} \pi_i \right)^2 v_i$$

$$= \frac{1}{2} \sum_{i=1}^{N} \sum_{j=1}^{N} \left(\frac{y_i}{\pi_i} - \frac{y_j}{\pi_j} \right)^2 \pi_i^2 \pi_j^2$$

$$\times \left[\frac{v_i t_j + v_j t_i}{\sum_{k=1}^{N} \pi_k^2 t_k} - t_i t_j \frac{\sum_{k=1}^{N} \pi_k^2 v_k}{(\sum_{k=1}^{N} \pi_k^2 t_k)^2} \right] \tag{17.13}$$

Comparing (17.13) with (17.8), we can see that goodness of fit of (16.30) to $E(\hat{Y} - Y)^2$ will depend on how well the coefficients

$$\delta_{ij} = \pi_i \pi_j - \pi_{ij} \qquad 1 \le i \ne j \le N \qquad (17.14)$$

are approximated by

$$\bar{\delta}_{ij} = \pi_i^2 \pi_j^2 \left[\frac{v_i t_j + v_j t_i}{\sum_{k=1}^N \pi_k^2 t_k} - t_i t_j \frac{\sum_{k=1}^N \pi_k^2 v_k}{(\sum_{k=1}^N \pi_k^2 t_k)^2} \right] \qquad 1 \le i \ne j \le N \qquad (17.15)$$

As may be easily seen, formula (17.13) will be tight in the sense introduced in Chapter 16 if the approximations $\bar{\delta}_{ij}$ are tight in the sense that

$$\sum_{j \ne i}^{N} \bar{\delta}_{ij} = \sum_{j \ne i}^{N} \delta_{ij} = \pi_i(1 - \pi_i) \qquad 1 \le i \le N \qquad (17.16)$$

[See Eq. (3.12) and the text following Eq. (7.28).] Actually, if (17.16) holds, then the coefficient of y_i^2 in (17.13) equals $(1 - \pi_i)/\pi_i = E(w_i - 1)^2$. Property (16.18) is also satisfied, since (17.13) vanishes for $y_i = x_i$, as is obvious from

$$\frac{x_i}{\pi_i} - \frac{x_j}{\pi_j} = \frac{X}{n} - \frac{X}{n} = 0 \qquad 1 \le i,j \le N$$

Our search for the approximate formula $E(\hat{Y} - Y)^2$ for the variance of the simple estimate $\hat{Y}$ will proceed as follows: First, being inspired by an asymptotic theory, we suggest $\bar{\delta}_{ij}$ in the form (17.15), i.e., we choose t_i and v_i. Then, in order to improve precision and to remove some misleading features, we adjust $\bar{\delta}_{ij}$ so as to achieve tightness or to approach tightness sufficiently well. A typical misleading feature can be, for example, that the approximate formula may give consistently larger values for one method than for another method. This is never true for exact formulas, but may be true for approximate formulas, even if they are asymptotically valid, as we shall see below. Tightness may be achieved either by modifying both t_i and v_i or by fixing t_i and replacing v_i by solutions of (16.32) which may for the simple estimate and $\pi_i = nx_i/X$ be rewritten as follows:

$$v_i = \frac{1 - \pi_i}{\pi_i}\left[1 - 2\frac{\pi_i^2 t_i}{\sum_{k=1}^N \pi_k^2 t_k} + \frac{\pi_i^2 t_i^2 / v_i}{(\sum_{k=1}^N \pi_k^2 t_k)^2} \sum_{k=1}^N \pi_k^2 v_k\right]^{-1}$$

$$1 \leq i \leq N \qquad (17.17)$$

Our first attempt to locate t_i and v_i will be based on the asymptotic results for $\delta_{ij} = \pi_i\pi_j - \pi_{ij}$ obtained for N and n both tending to infinity.

As for successive sampling, Theorem 9.9 provides

$$\bar{\delta}_{ij} = (1 - \pi_i)(1 - \pi_j)\left\{\frac{\pi_i\alpha_j + \pi_j\alpha_i}{\sum_{k=1}^N \alpha_k(1 - \pi_k)} - \alpha_i\alpha_j\frac{\sum_{k=1}^N \pi_k(1 - \pi_k)}{\left[\sum_{k=1}^N \alpha_k(1 - \pi_k)\right]^2}\right\} \qquad 1 \leq i \neq j \leq N \qquad (17.18)$$

Comparing (17.18) to (17.15), we can see that the right-hand sides coincide for

$$t_i = \frac{\alpha_i(1 - \pi_i)}{\pi_i^2} \qquad 1 \leq i \leq N \qquad (17.19)$$

and

$$v_i = \frac{1 - \pi_i}{\pi_i} \qquad 1 \leq i \leq N \qquad (17.20)$$

A tight formula could be obtained by replacing v_i of (17.15) by solutions of (17.17), where t_i is defined by (17.19). However, we shall not attempt it.

As for rejective sampling, a comparison of right-hand sides of (17.15) and (7.27) provides

$$t_i = v_i = \frac{1 - \pi_i}{\pi_i} \qquad 1 \leq i \leq N \qquad (17.21)$$

Alternatively, we may use (7.28), which yields

$$t_i = v_i = \frac{1 - \lambda_i}{\pi_i} \qquad 1 \leq i \leq N \qquad (17.22)$$

where $1 - \lambda_i$ are the solutions of (3.24). The formula based on (17.22) is tight. Correspondingly, (17.7) turns into (3.24) if we substitute $(1 - \lambda_i)/\pi_i$ for v_i as well as for t_i.

In view of Theorem 8.2, (17.21) and (17.22) are applicable also for Sampford-Durbin sampling.

Next we shall utilize asymptotic results for $\delta_{ij} = \pi_i\pi_j - \pi_{ij}$ under n fixed and $N \to \infty$. As we can see from (7.39), (8.19), and (9.55), we can write

$$\bar{\delta}_{ij} = \frac{\pi_i(1 - \lambda_i)\pi_j(1 - \lambda_j)}{\sum_{k=1}^{N} \pi_k(1 - \lambda_k)} \times [1 - a_n(\pi_i - \bar{\bar{\pi}})(\pi_j - \bar{\bar{\pi}})] \qquad 1 \le i \ne j \le N \tag{17.23}$$

where $1 - \lambda_i$ are the solutions of (3.24), $\bar{\bar{\pi}} = \Sigma_{i=1}^{N} \pi_i^2(1 - \lambda_i)/\Sigma_{i=1}^{N} \pi_i(1 - \lambda_i)$, and

$$a_n = \begin{cases} \frac{1}{4}\frac{(n-1)(n+5)}{n^2} & \text{for successive sampling} \\ 2\frac{n-1}{n^2} & \text{for rejective sampling} \\ \frac{n-1}{n^2} & \text{for Sampford-Durbin sampling} \end{cases} \tag{17.24}$$

Comparing the right-hand sides of (17.23) and (17.15), we can see that they coincide for

$$t_i = \frac{(1 - \lambda_i)[1 + \sqrt{a_n}(\pi_i - \bar{\bar{\pi}})]}{\pi_i} \qquad 1 \le i \le N \tag{17.25}$$

and

$$v_i = \frac{1 - \lambda_i}{\pi_i} \qquad 1 \le i \le N \tag{17.26}$$

In order to see it, we must note that, in view of the definition of $\bar{\bar{\pi}}$,

$$\sum_{i=1}^{N} \pi_i^2 t_i = \sum_{i=1}^{N} \pi_i(1 - \lambda_i)[1 + \sqrt{a_n}(\pi_i - \bar{\bar{\pi}})]$$

$$= \sum_{i=1}^{N} \pi_i(1 - \lambda_i) = \sum_{i=1}^{N} \pi_i^2 v_i \tag{17.27}$$

Of course, since the t_i terms play the role of weights, we can multiply them by any constant number.

For n large and the π_i terms small, both theories give concordant results, as it should be. As for the weights t_i, the conformity of (17.25) and (17.19) follows from (9.9) and from $\sqrt{a_n} \to 1/2$ for $n \to \infty$, which holds in view of the first equation of (17.24). Similarly for rejective and Sampford-Durbin sampling, (17.25) is concordant with (17.22) in view of $\sqrt{a_n} \to 0$ for $n \to \infty$.

Our further investigations will be based on (17.25) for the following reasons: First, it is able to distinguish between all the three methods; second, it gives more precise results unless the π_i terms are unusually large. This somewhat unexpected success of the theory for n fixed is due to the fact that we were able to establish not only the main term but also the next one.

Of course, we could obtain the next term also under $n \to \infty$, as we succeeded for rejective sampling in Chapter 14 in formula (14.13). However, it would need diabolical patience and, moreover, in order to obtain simple and comparable results we would have to assume that the π_i terms are small, which would probably bring us back to (17.25).

Let us now analyze the regression coefficient $\bar{\beta}$ defined by means of weights t_i from (17.25). In view of (16.31) and since $\pi_i = nx_i/X$, we can write

$$\bar{\beta} = \frac{\sum_{i=1}^{N} y_i x_i t_i}{\sum_{i=1}^{N} x_i^2 t_i} = \frac{\sum_{i=1}^{N} y_i(1 - \lambda_i)[1 + \sqrt{a_n}(\pi_i - \bar{\bar{\pi}})]}{\sum_{i=1}^{N} x_i(1 - \lambda_i)[1 + \sqrt{a_n}(\pi_i - \bar{\bar{\pi}})]}$$

$$= \frac{\sum_{i=1}^{N} y_i(1 - \lambda_i)[1 + \sqrt{a_n}(\pi_i - \bar{\bar{\pi}})]}{\sum_{i=1}^{N} x_i(1 - \lambda_i)}$$

$$= \varphi + \sqrt{a_n}\, \frac{\sum_{i=1}^{N} y_i(\pi_i - \bar{\bar{\pi}})(1 - \lambda_i)}{\sum_{i=1}^{N} x_i(1 - \lambda_i)} \tag{17.28}$$

where

$$\varphi = \frac{\sum_{i=1}^{N} y_i(1 - \lambda_i)}{\sum_{i=1}^{N} x_i(1 - \lambda_i)} \tag{17.29}$$

Now φ is the least-squares estimate of the regression coefficient of y_i on x_i with weights $(1 - \lambda_i)/\pi_i$, so that

$$\sum_{i=1}^{N} (y_i - \bar{\beta} x_i)^2 \frac{1 - \lambda_i}{\pi_i} = \sum_{i=1}^{N} (y_i - \varphi x_i)^2 \frac{1 - \lambda_i}{\pi_i} + (\bar{\beta} - \varphi)^2 \sum_{i=1}^{N} \frac{x_i^2}{\pi_i}(1 - \lambda_i) \tag{17.30}$$

Making use of (17.28) we can further write

$$(\bar{\beta} - \varphi)^2 \sum_{i=1}^{N} \frac{x_i^2}{\pi_i}(1 - \lambda_i) = a_n \frac{\left[\sum_{i=1}^{N} y_i(\pi_i - \bar{\bar{\pi}})(1 - \lambda_i)\right]^2}{\sum_{i=1}^{N} \pi_i(1 - \lambda_i)} \tag{17.31}$$

Next

$$\begin{aligned}\sum_{i=1}^{N} y_i(\pi_i - \bar{\bar{\pi}})(1 - \lambda_i) &= \sum_{i=1}^{N} \frac{y_i}{\pi_i}(\pi_i - \bar{\bar{\pi}})\pi_i(1 - \lambda_i) \\ &= \sum_{i=1}^{N} \left(\frac{y_i}{\pi_i} - \varphi \frac{X}{n}\right)(\pi_i - \bar{\bar{\pi}})\pi_i(1 - \lambda_i) \\ &= \zeta \left[\sum_{i=1}^{N} \left(\frac{y_i}{\pi_i} - \varphi \frac{X}{n}\right)^2 \pi_i(1 - \lambda_i)\right]^{1/2} \\ &\quad \times \left[\sum_{i=1}^{N} (\pi_i - \bar{\bar{\pi}})^2 \pi_i(1 - \lambda_i)\right]^{1/2} \\ &= \zeta \left[\sum_{i=1}^{N} (y_i - \varphi x_i)^2 \frac{1 - \lambda_i}{\pi_i}\right]^{1/2} \\ &\quad \times \left[\sum_{i=1}^{N} (\pi_i - \bar{\bar{\pi}})^2 \pi_i(1 - \lambda_i)\right]^{1/2}\end{aligned} \tag{17.32}$$

where, obviously, ζ is the weighted correlation coefficient of y_i/π_i and π_i with weights $\pi_i(1 - \lambda_i)$. Equivalently, ζ is the weighted correlation coefficient of y_i/x_i and x_i with weights $x_i(1 - \lambda_i)$. Substituting (17.32) into (17.31), we have

$$(\bar{\beta} - \varphi)^2 \sum_{i=1}^{N} \frac{x_i^2}{\pi_i} (1 - \lambda_i) = a_n\zeta^2 \left[\sum_{i=1}^{N} (y_i - \varphi x_i)^2 \frac{1 - \lambda_i}{\pi_i} \right] \delta^2 \tag{17.33}$$

where δ^2 is the weighted variance of π_i with weights $\pi_i(1 - \lambda_i)$:

$$\delta^2 = \frac{\sum_{i=1}^{N} (\pi_i - \bar{\bar{\pi}})^2 \pi_i(1 - \lambda_i)}{\sum_{i=1}^{N} \pi_i(1 - \lambda_i)} \tag{17.34}$$

Finally, inserting (17.33) into (17.30), we obtain

$$\sum_{i=1}^{N} (y_i - \bar{\beta}x_i)^2 \frac{1 - \lambda_i}{\pi_i} = \left[\sum_{i=1}^{N} (y_i - \varphi x_i)^2 \frac{1 - \lambda_i}{\pi_i} \right] (1 + a_n\zeta^2\delta^2) \tag{17.35}$$

In formula (17.35) the differences between the three sampling methods are reflected only in different values of a_n according to (17.24). This approximate formula is misleading in suggesting that a method with smallest a_n gives smallest variance uniformly in $(y_1, \ldots, y_N)$. According to (17.24) the smallest a_n is associated for all n with Sampford-Durbin sampling. Consequently, Sampford-Durbin sampling is best according to formula (17.35). The relation between successive and rejective sampling depends on n: for $n = 2$ a_n is smaller for successive sampling, for $n = 3$ there is a tie, and for $n > 3$, a_n is smaller for rejective sampling. Some selected values of a_n are given in Table 17.1.

In order to remove from (17.35) the phenomenon of uniform dominance, we have to make a correction toward tightness-- of course, without losing the attained degree of precision. Given the weights t_i from (17.25), a tight fòrmula is provided for v_i satisfying (17.17). An approximate solution of (17.17) may be obtained as follows: We pretend that the solution is a constant multiple of $(1 - \lambda_i)/\pi_i$:

TABLE 17.1 Values of a_n defined in Eq. (17.24)

n	Sampling		
	Successive	Rejective	Sampford-Durbin
2	.4375	.5000	.2500
3	.4444	.4444	.2222
4	.4219	.3750	.1875
5	.4000	.3200	.1600
∞	.2500	.0000	.0000

$$v_i = \frac{1 - \lambda_i}{\pi_i} c \tag{17.36}$$

Then, in view of (3.24) and (17.17),

$$c = \frac{(1 - \pi_i)/(1 - \lambda_i)}{(1 - \pi_i)/\pi_i v_i} = \frac{\Sigma_{i=1}^N (1 - \pi_i)/(1 - \lambda_i)}{\Sigma_{i=1}^N (1 - \pi_i)/\pi_i v_i}$$

$$= \frac{\Sigma_{i=1}^N \left[1 - \dfrac{\pi_i(1 - \lambda_i)}{\Sigma_{k=1}^N \pi_k(1 - \lambda_k)}\right]}{\Sigma_{i=1}^N \left[1 - 2\dfrac{\pi_i^2 t_i}{\Sigma_{k=1}^N \pi_k^2 t_k} + \dfrac{\pi_i^2 t_i^2/v_i}{(\Sigma_{k=1}^N \pi_k^2 t_k)^2} \Sigma_{k=1}^N \pi_k^2 v_k\right]}$$

Now the sum in the numerator equals N - 1, while the sum in the denominator equals $N - 1 + V_*^2$, where V_* is the weighted coefficient of variation of t_i/v_i with weights $\pi_i^2 v_i$. In order to see the last-stated fact, we only have to apply (1.21) to prove

$$\frac{\Sigma_{i=1}^N (\pi_i^2 t_i^2/v_i) \Sigma_{k=1}^N \pi_k^2 v_k}{(\Sigma_{k=1}^N \pi_k^2 t_k)^2} = 1 + V_*^2$$

However, (17.25) and (17.36) entail that $t_i/v_i = [1 + \sqrt{a_n}(\pi_i - \bar{\bar{\pi}})]/c$ and $\pi_i^2 v_i = \pi_i(1 - \lambda_i)c$ so that $V_*^2 = a_n\delta^2$, δ^2 being defined by (17.34).

Thus

$$c = \frac{N - 1}{N - 1 + a_n \delta^2}$$

and

$$v_i = \frac{1 - \lambda_i}{\pi_i} \frac{N - 1}{N - 1 + a_n \delta^2} \tag{17.37}$$

Replacing in (17.35) the weights $(1 - \lambda_i)/\pi_i$ by (17.37), we obtain formula

$$\sum_{i=1}^{N} (y_i - \bar{\beta} x_i)^2 \frac{1 - \lambda_i}{\pi_i} \frac{N - 1}{N - 1 + a_n \delta^2}$$

$$= \sum_{i=1}^{N} (y_i - \varphi x_i)^2 \frac{1 - \lambda_i}{\pi_i} (1 + a_n \zeta^2 \delta^2) \left(1 + a_n \frac{\delta^2}{N - 1}\right)^{-1} \tag{17.38}$$

The first service provided by formula (17.38) was the conclusion that the efficiency of the three sampling methods is largely governed by ζ, the correlation coefficient of y_i/x_i and x_i. The next service will consist in giving us an idea of how big the differences in variance may be expected to be.

Consider two methods P' and P" and denote the corresponding constants a_n by a_n' and a_n''. Then the ratio of approximate variances will equal

$$\frac{\text{var}(P')}{\text{var}(P'')} = \frac{1 + a_n'[\zeta^2 - 1/(N - 1)]\delta^2}{1 + a_n''[\zeta^2 - 1/(N - 1)]\delta^2} \tag{17.39}$$

Now we have, in view of (17.24), the inequality $a_n \le 0.5$, and almost without exception $\delta^2 = 0.005$ and $|\zeta^2 - 1/(N - 1)| = .4$. Consequently the variance ratio will lie within the limits

$$1 \pm .001$$

and usually will be in much narrower bounds. Consequently, our second conclusion drawn from (17.38) is that all the methods are practically equally efficient.

If $a_n' < a_n''$, then (17.39) entails

$$\frac{1 + Na_n'\delta^2/(N-1)}{1 + Na_n''\delta^2/(N-1)} \le \frac{\text{var}(P')}{\text{var}(P'')} \ne \frac{1 - a_n'\delta^2/(N-1)}{1 - a_n''\delta^2/(N-1)} \tag{17.40}$$

where the extremes correspond to $\zeta^2 = 1$ and $\zeta^2 = 0$. Consequently, the maximum possible relative reduction of variance achieved by var(P') over var(P") is N times larger than the same achieved by var(P") over var(P'). This explains why a very small modification of the approximate formulas may cause var(P') ≥ var(P") to hold for every $(y_1, \ldots, y_N)$.

Though much simpler than the Yates-Grundy formula, (17.38) is still too complicated for everyday use. Since the differences between the methods are negligible, we should suggest a simple formula serving all three of them. Such a formula may be provided by $t_i = 1/\pi_i$, which usually differ only little from (17.25) if the π_i terms are not too large. Since the weights may be multiplied by any fixed number, we may equivalently put $t_i = 1/x_i$. This yields

$$\bar{\beta} = \frac{\sum_{i=1}^N y_i x_i t_i}{\sum_{i=1}^N x_i^2 t_i} = \frac{\sum_{i=1}^N y_i}{\sum_{i=1}^N x_i} = \frac{Y}{X} \tag{17.41}$$

A small change in the weight t_i implies a still smaller change in $\bar{\beta}$, and an even more negligible change in $\Sigma_{i=1}^N (y_i - \bar{\beta}x_i)^2 v_i$ if the formula is tight. The solution of (17.17) for $t_i = 1/\pi_i$ may be approximately obtained by putting $t_i/v_i = c$ in the right-hand side. Then we may substitute

$$\frac{\pi_i^2 t_i^2/v_i}{(\sum_{k=1}^N \pi_k^2 t_k)^2} \sum_{k=1}^N \pi_k^2 v_k = \frac{\pi_i^2 t_i}{\sum_{k=1}^N \pi_k^2 t_k}$$

and (17.17) yields

$$v_i = \frac{1 - \pi_i}{\pi_i}\left(1 - \frac{\pi_i^2 t_i}{\sum_{k=1}^N \pi_k^2 t_k}\right)^{-1} = \frac{1 - \pi_i}{\pi_i}\left(1 - \frac{\pi_i}{n}\right)^{-1} \qquad 1 \le i \le N \tag{17.42}$$

In terms of the x_i terms, we have

$$v_i = \frac{X}{nx_i} \frac{X - nx_i}{X - x_i} \qquad 1 \le i \le N \tag{17.43}$$

The v_i terms given by (17.42) will be in the average very slightly larger than exact solutions of (17.17).

Inserting (17.41) and (17.43) into (16.30) we obtain

$$E(\hat{Y} - Y)^2 = \sum_{i=1}^{N} \left(y_i - \frac{Y}{X} x_i\right)^2 \frac{X}{nx_i} \frac{X - nx_i}{X - x_i} \tag{17.44}$$

If $x_i = 1$, the formula coincides with (17.11), and consequently is precise. If the model of linear regression passing through the origin is correct, the value given by (17.44) will oscillate around the exact value. If $\zeta^2 > 1/(N - 1)$, (17.44) will be a slight overestimate of the exact variance, with the possible exception of $n = 2$ and rejective and successive sampling.

One may conjecture that (17.44) is applicable also for randomized systematic sampling.

We recommend the formula even if it is not asymptotically valid for any sampling method under consideration. This is because for real populations a formula which is almost tight and approximately valid asymptotically may be better than another which is asymptotically valid but not tight--unless the underlying model $y_i = \beta x_i$ + random deviation is drastically wrong for the population.

PROBLEMS AND NOTES

17.1 Under Poisson sampling with inclusion probabilities $\pi_i > 0$, $1 \le i \le N$, the simple estimate $\hat{Y}$ has the following moments:

$$E\hat{Y} = Y$$

$$E(\hat{Y} - Y)^2 = \sum_{i=1}^{N} \frac{y_i^2(1 - \pi_i)}{\pi_i}$$

$$E(\hat{Y} - Y)^3 = \sum_{i=1}^{N} \frac{y_i^3(1 - \pi_i)(1 - 2\pi_i)}{\pi_i^2} \tag{17.45}$$

$$E(\hat{Y} - Y)^4 = \sum_{i=1}^{N} \frac{y_i^4(1 - \pi_i)[1 - 6\pi_i(1 - \pi_i)]}{\pi_i^3} + 3[E(\hat{Y} - Y)^2]^2 \tag{17.45}$$

cont'd

17.2 Under Poisson sampling with inclusion probabilities $\pi_i > 0$, $1 \leq i \leq N$, the simple estimates $\hat{Y}$, $\hat{X}$, $\hat{U}$, $\hat{Z}$ have the following mixed moments:

$$\mathrm{Cov}(\hat{Y},\hat{X}) = \sum_{i=1}^{N} \frac{y_i x_i(1 - \pi_i)}{\pi_i}$$

$$E[(\hat{Y} - Y)(\hat{X} - X)(\hat{U} - U)] = \sum_{i=1}^{N} \frac{y_i x_i u_i(1 - \pi_i)(1 - 2\pi_i)}{\pi_i^2}$$

$$\begin{aligned} &E[(\hat{Y} - Y)(\hat{X} - X)(\hat{U} - U)(\hat{Z} - Z)] \\ &\quad = \sum_{i=1}^{N} \frac{y_i x_i u_i z_i(1 - \pi_i)[1 - 6\pi_i(1 - \pi_i)]}{\pi_i^3} + \mathrm{Cov}(\hat{Y},\hat{Z})\mathrm{Cov}(\hat{U},\hat{Z}) \\ &\quad + \mathrm{Cov}(\hat{Y},\hat{U})\mathrm{Cov}(\hat{X},\hat{Z}) + \mathrm{Cov}(\hat{Y},\hat{Z})\mathrm{Cov}(\hat{X},\hat{U}) \end{aligned}$$

17.3 Under successive, rejective, and Sampford-Durbin sampling with $\pi_i = nx_i/X$, $1 \leq i \leq N$, we may put

$$\bar{E}[(\hat{Y} - Y)(\hat{Z} - Z)] = \sum_{i=1}^{N} \left(y_i - \frac{Y}{X} x_i\right)\left(z_i - \frac{Z}{X} x_i\right)\left(\frac{X}{nx_i}\right)\frac{X - nx_i}{X - x_i}$$

17.4 Under simple random sampling

$$E(\hat{X} - X)^3 = \left(\frac{N}{n}\right)^2 \frac{N - n}{N - 1}\,\frac{N - 2n}{N - 2} \sum_{i=1}^{N} (x_i - \bar{X})^3$$

and

$$\begin{aligned} E(\hat{X} - X)^4 &= 3\,\frac{N - 1}{N + 1}(\mathrm{Var}\ \hat{Y})^2 + \frac{N^4(N - n)(N + 1)}{n^3(N - 1)(N - 2)(N - 3)} \\ &\quad \times \left[1 - 6\,\frac{n(N - n)}{N(n + 1)}\right]\left\{\sum_{i=1}^{N} (x_i - \bar{X})^4 \right. \\ &\quad \left. - 3\,\frac{N - 1}{N(N + 1)}\left[\sum_{i=1}^{N} (x_i - \bar{X})^2\right]^2\right\} \end{aligned}$$

17.5 If y_i indicates the presence of a property A with relative frequency P_A, then under simple random sampling

$$\hat{Y} = Np_A$$

where p_A is the relative frequency in the sample, and

$$\text{Var } \hat{Y} = P_A(1 - P_A) \frac{N^2}{n} \frac{N - n}{N - 1}$$

17.6 Let $\hat{Y}$ be the simple estimate and P and Q two probability samplings of fixed size n and with coinciding inclusion probabilities π_i, $1 \leq i \leq N$. Show that

$$\text{Var}(\hat{Y}|P) \leq \text{Var}(\hat{Y}|Q)$$

cannot hold uniformly for all $(y_1,\ldots,y_N)$, unless $\text{Var}(\hat{Y}|P) = \text{Var}(\hat{Y}|Q)$ identically. Here $\text{Var}(\hat{Y}|P)$ denotes the variance of $\hat{Y}$ under P.

17.7 *Admissibility* Given any fixed-size probability sampling, the simple estimate is admissible in the class of all Lebesgue measurable estimates (including nonlinear and biased) (Joshi, 1965). For definition of admissibility, see Chapter 4.

18

OPTIMUM SAMPLING-ESTIMATING STRATEGIES

Assume that the genesis of the y values may be described by a superpopulation model. Most simple and also most useful superpopulation models are the *regression models*, special cases of which have already been considered in Part I (see Example 2.3). We shall say that the y_i terms follow a regression model if they may be represented as

$$y_i = \sum_{k=1}^{m} \beta_k x_{ki} + \lambda b_i e_i \qquad 1 \le i \le N \tag{18.1}$$

where $x_{1i}, \ldots, x_{mi}, b_i$ are some known numbers, $\beta_1, \ldots, \beta_m, \lambda > 0$ are unknown parameters, and e_i are now normed residuals, i.e., $Me_i = 0$ and $Me_i^2 = 1$, $1 \le i \le N$, where M denotes the expectation relative to the superpopulation model. Unless stated otherwise, we shall assume that the residuals are *uncorrelated*. Typically, x_{ki} will be values of some variables x_k ascertained a priori and $b_i = f(x_{1i},\ldots,x_{mi})$ for some function f. As in Chapter 16, we shall call the variables x_k *regressors*.

The Bayesian sampling-estimating program may be formulated as follows:

1. We fix a sample s and find an estimate $\hat{Y} = \hat{Y}_s$ for which

$$M\hat{Y}_s = MY \qquad \text{and} \qquad M(\hat{Y}_s - Y)^2 = \text{minimum} \tag{18.2}$$

2. Denoting the minimum value of $M(\hat{Y}_s - Y)^2$ by B_s and the costs associated with a sample s by C_s, we select a sample s that minimizes B_s under the side condition $C_s \le C$, where C is a certain cost limit.

THEOREM 18.1 For a given sample s the expression $M(\hat{Y} - Y)^2$ is minimized under the side condition $M\hat{Y} = MY$ by

$$\hat{Y}_B = y + \sum_{j=1}^{m} (X_j - x_j)\hat{\beta}_j \tag{18.3}$$

where $y = \Sigma_{i \in s} y_i$ and $\hat{\beta}_1, \ldots, \hat{\beta}_m$ are the best linear unbiased estimates for $\beta_1, \ldots, \beta_m$ based on $\{y_i, i \in s\}$. Introducing the matrix

$$G = \left\{\sum_{i \in s} \frac{x_{ji}x_{ki}}{b_i^2}\right\}_{j,k=1}^{m} \tag{18.4}$$

and its inverse

$$H = G^{-1} = \{h_{j,k}\}_{j,k=1}^{m} \tag{18.5}$$

assuming that Det $G > 0$, we have

$$\hat{\beta} = (\hat{\beta}_1, \ldots, \hat{\beta}_m)' = H\left(\sum_{i \in s} \frac{y_i x_{1i}}{b_i^2}, \ldots, \sum_{i \in s} \frac{y_i x_{mi}}{b_i^2}\right)' \tag{18.6}$$

and the minimum value of $M(\hat{Y} - Y)^2$ equals

$$M(\hat{Y}_B - Y)^2 = \lambda^2\left[\sum_{j=1}^{m} \sum_{k=1}^{m} (X_j - x_j)(X_k - x_k)h_{jk} + \sum_{i \in r} b_i^2\right] \tag{18.7}$$

where $x_j = \Sigma_{i \in s} x_{ji}$, $X_j = \Sigma_{i=1}^{N} x_{ji}$, and $r = S - s$.

Proof. We have, in view of $M\hat{Y} = MY$,

$$\begin{aligned} M(\hat{Y} - Y)^2 &= M\left(\hat{Y} - y - \sum_{i \in r} y_i\right)^2 \\ &= M\left[\hat{Y} - y - \sum_{j=1}^{m} (X_j - x_j)\beta_j\right]^2 + \lambda^2 \sum_{i \in r} b_i^2 \end{aligned} \tag{18.8}$$

Thus the best $\hat{Y}$ has the form (18.3). The rest follows from the general theory of regression (see, e.g., Rao, 1965). Q.E.D.

THEOREM 18.2 Assuming that Det $G > 0$, $n > m$, and putting

$$\hat{\lambda}^2 = \frac{1}{n - m} \sum_{i \in s} \frac{[y_i - \Sigma_{j=1}^{m} \hat{\beta}_j x_{ji}]^2}{b_i^2} \tag{18.9}$$

where $\hat{\beta}$ is given by (18.6), we have

$$M\hat{\lambda}^2 = \lambda^2 \tag{18.10}$$

Furthermore, if the vector $(y_1,\ldots,y_N)$ is normal, then $\hat{\beta}$ and $\hat{\lambda}^2$ are the best unbiased estimates for β and λ^2, they are mutually independent, and the ratio

$$t = \frac{\hat{Y} - Y}{\sqrt{v}} \tag{18.11}$$

where

$$v = \hat{\lambda}^2 \left[\sum_{j=1}^{m} \sum_{k=1}^{m} (X_j - x_j)(X_k - x_k)h_{jk} + \sum_{i \in r} b_i^2 \right] \tag{18.12}$$

has the Student distribution with n - m degrees of freedom.

Proof. Again, it is a well-known result of the theory of regression and of best unbiased estimation (Rao, 1965).

Example 18.1 Considering the model

$$y_i = \beta x_i + \lambda e_i \qquad 1 \le i \le N \tag{18.13}$$

the above theorem yields

$$\hat{Y}_B = y + (X - x) \frac{\Sigma_{i \in s} y_i x_i}{\Sigma_{i \in s} x_i^2} \tag{18.14}$$

and

$$v = \frac{1}{n - 1} \sum_{i \in s} \left(y_i - \frac{\Sigma_{j \in s} y_j x_j}{\Sigma_{j \in s} x_j^2} x_i \right)^2 \left[\frac{(X - x)^2}{\Sigma_{i \in s} x_i^2} + (N - n) \right] \tag{18.15}$$

If all units are equally expensive, i.e., $C_s = c(\text{size } s)$, then the best sample of given size will consist of n units bearing the largest values x_i. If the costs associated with various units vary, we shall have to compromise between the requirements that the units included in the sample should be "cheap," on one hand, and possess large x_i, on the other hand.

Other illustrations were given in Examples 2.1 and 2.3 and Problem 2.1.

However impressive they may seem, the results of the above theorems are not convincing. We cannot help asking what happens if

1. The regression model (18.1) is oversimplified; or
2. The y_i terms are correlated.

Example 18.2 Assume that the true regression is

$$y_i = \alpha + \beta x_i + \lambda e_i \qquad 1 \le i \le N \tag{18.16}$$

whereas we assumed (18.13) and employed estimate (18.14). Then

$$M(\hat{Y}_B - Y) = \alpha\left[\frac{x(X - x)}{\Sigma_{i\in s}\, x_i^2} - (N - n)\right] \tag{18.17}$$

and the unbiasedness (with respect to the superpopulation) is lost.

Example 18.3 For the model

$$y_i = \alpha + \lambda e_i \qquad 1 \le i \le N \tag{18.18}$$

the error estimate (18.15) becomes

$$v = \frac{N(N - n)}{n(n - 1)} \sum_{i\in s} (y_i - \bar{y})^2 \tag{18.19}$$

where $\bar{y}$ is the sample average $\bar{y} = y/n$. However, if the y_i terms are correlated, we have for $Y = N\bar{y}$,

$$M(\hat{Y}_B - Y)^2 = \lambda^2\left[\frac{N(N - n)}{n} + \left(\frac{N}{n} - 1\right)^2 \sum_{i\ne j}^{s}\sum^{s} \zeta_{ij} - 2\left(\frac{N}{r} - 1\right)\sum_{i\in s}\sum_{j\in r} \zeta_{ij} + \sum_{i\ne j}^{r}\sum^{r} \zeta_{ij}\right] \tag{18.20}$$

where $\Sigma^s_{i\ne j}\Sigma^s$ extends over all pairs from the sample s, to be compared with

$$M(v) = \lambda^2 \left[\frac{N(N - n)}{n} - \frac{N(N - n)}{n^2(n - 1)} \sum_{i \neq j}^{s} \sum^{s} \zeta_{ij} \right] \tag{18.21}$$

where ζ_{ij} is the correlation coefficient of y_i and y_j.

Thus we do not obtain $M(v) = M(\hat{Y}_B - Y)^2$, and, for example, if $\zeta_{ij} > 0$ prevails in the sample, v will be a heavy underestimate.

We shall look for protection against the possible imperfections of the regression model. Of course this protection will not be gained for nothing. We shall be forced to use estimates and samples that are not best ones under strict validity of the regression model (18.1).

The robust sampling-estimating program may be formulated as follows:

1. We consider a probability sampling $\{P(s), s \subset S\}$ and denote the expectation with respect to it by $E(\cdot)$, i.e., $E(\cdot) = E(\cdot)P(s)$. Then we find an estimate which satisfies

$$E\hat{Y} = Y \tag{18.22}$$

and

$$EM(\hat{Y} - Y)^2 = \text{minimum} \tag{18.23}$$

2. Denoting for a while the expectation relative to the sampling design $P(\cdot)$ by E_p, the minimum value of $E_p M(\hat{Y} - Y)^2$ by B_p, and the costs associated with a sample s by C_s, we shall choose such a sampling design P that minimizes B_p under the side condition

$$C_p = E_p(C_s) \leq C \tag{18.24}$$

where C is some cost limit.

3. From among the sampling designs satisfying 2 we shall try to choose those for which the Bayesian error estimate and also the distribution of the Student ratio (18.11) are robust.

Remark 18.1 If the satisfaction of (18.22) is difficult, we may replace it by a weaker requirement, namely that the bias is negligible in ordinary situations.

THEOREM 18.3 Assume that the regression model 18.1 holds and that the variable part of expected costs equals

$$C_P = \sum_{i=1}^{N} c_i \pi_i \tag{18.25}$$

where the π_i terms are the probabilities of inclusion associated with the sampling design P.

Then the sampling-estimating strategy (P,w) such that

1° $\quad \pi_i = \varkappa \dfrac{b_i}{\sqrt{c_i}} \qquad 1 \le i \le N$ (18.26)

2° $\quad C_p = C$ (18.27)

3° $\quad \hat{Y} = \Sigma_{i=1}^{N} y_i w_i$ is the simple estimate

4° $\quad \hat{Y}$ is representative with respect to the values $x_{j1}, \ldots, x_{jN}$ for all $j = 1, \ldots, m$

is best in the sense that it minimizes $EM(\hat{Y} - Y)^2$ under $E\hat{Y} = Y$ and $\Sigma_{i=1}^{N} c_i \pi_i \le C$.

Proof. Denoting $\hat{Y} = \Sigma_{i=1}^{N} y_i w_i$, where $w_i = 0$ if $i \notin s$, we have

$$M(\hat{Y} - Y)^2 = \left[\sum_{i=1}^{N} \sum_{j=1}^{m} \beta_j x_{ji}(w_i - 1)\right]^2 + \lambda^2 \sum_{i=1}^{N} b_i^2 (w_i - 1)^2$$
$$\ge \lambda^2 \sum_{i=1}^{N} b_i^2 (w_i - 1)^2 \tag{18.28}$$

where the case of equality occurs if and only if 4° is satisfied, i.e., if (16.19) holds.

Thus

$$EM(\hat{Y} - Y)^2 \ge \lambda^2 \sum_{i=1}^{N} b_i^2 E(w_i - 1)^2 \tag{18.29}$$

Now unbiasedness in the sense of $E\hat{Y} = Y$ is equivalent to

$$Ew_i = 1 \qquad 1 \le i \le N \tag{18.30}$$

Since $w_i = 0$ for $i \notin s$, we have

$$\begin{aligned} E(w_i - 1)^2 &= \pi_i E[(w_i - 1)^2 | s \ni i] + 1 - \pi_i \\ &\ge \pi_i [E(w_i - 1 | s \ni i)]^2 + 1 - \pi_i \end{aligned} \tag{18.31}$$

where the equality occurs if and only if

$$w_i = E(w_i | s \ni i) \qquad \text{if } s \ni i \tag{18.32}$$

However,

$$Ew_i = \pi_i E(w_i | s \ni i) = 1 \tag{18.33}$$

Combining (18.32) and (18.33), we have

$$w_i = \frac{1}{\pi_i} \qquad \text{if } s \ni i \tag{18.34}$$

Thus

$$E(w_i - 1)^2 \ge \pi_i \left(\frac{1}{\pi_i} - 1\right)^2 + 1 - \pi_i = \frac{1}{\pi_i} - 1$$

and the equality is achieved if (18.34) holds, i.e., if 3° is satisfied.

Consequently, if 3° and 4° are granted, we have to minimize

$$\begin{aligned} EM(\hat{Y} - Y)^2 &= \lambda^2 \sum_{i=1}^{N} b_i^2 \left(\frac{1}{\pi_i} - 1\right) \\ &= \lambda^2 \sum_{i=1}^{N} \frac{b_i^2}{\pi_i} - \lambda^2 \sum_{i=1}^{N} b_i^2 \end{aligned} \tag{18.35}$$

under

$$\sum_{i=1}^{N} c_i \pi_i \le C \tag{18.36}$$

Using the result of Problem 1.8, we have

$$\sum_{i=1}^{N} \frac{b_i^2}{\pi_i} = \left(\sum_{i=1}^{N} \frac{b_i^2}{\pi_i}\right) \frac{\sum_{i=1}^{N} c_i \pi_i}{C} \geq \frac{\left(\sum_{i=1}^{N} b_i \sqrt{c_i}\right)^2}{C} \tag{18.37}$$

where the case of equality occurs if and only if 1° and 2° are satisfied. Q.E.D.

Remark 18.2 As may be seen from (18.35) and (18.37) the lower bound for $EM(\hat{Y} - Y)^2$ equals

$$EM(\hat{Y} - Y)^2 = \lambda^2 \left[\frac{\left(\sum_{i=1}^{N} b_i \sqrt{c_i}\right)^2}{C} - \sum_{i=1}^{N} b_i^2\right] \tag{18.38}$$

This lower bound may be exceeded only at the expense of the loss of unbiasedness $E\hat{Y} = Y$. We can also see that small deviations from unbiasedness in the sense that Ew_i differs from 1 only slightly can diminish $EM(\hat{Y} - Y)^2$ only very little. Thus an essential decrement in $EM(\hat{Y} - Y)^2$ can be achieved only for heavily biased estimates. In a given situation the best solution should be found somewhere between Theorems 18.1 and 18.3, depending on how firmly we can believe in the regression model (18.1).

Remark 18.3 Theorem 18.3 solves only points 1 and 2 of the above robust sampling-estimating program. On the other hand, it prescribes the sampling design only as far as the π_i terms are concerned. It leaves the π_{ij} terms as well as other aspects of P open. These may be chosen so as to satisfy point 3°.

Remark 18.4 The basic assumption in Theorem 18.3 is that the residuals are *uncorrelated*. This assumption may be removed if we assume that the sampling-estimating strategy makes it possible to utilize the approximation (16.27) for $E(\hat{Y} - Y)^2$. Then provided only that $\bar{\beta}_k$ are good estimates of β_k, we can write

$$M\left(y_i - \sum_{k=1}^{m} \bar{\beta}_k x_{ki}\right)^2 \doteq M\left(y_i - \sum_{k=1}^{m} \beta_k x_{ki}\right)^2$$

$$= \lambda^2 b_i^2 \qquad 1 \leq i \leq N \tag{18.39}$$

and

$$\mathrm{ME}(\hat{Y} - Y)^2 \doteq \lambda^2 \sum b_i^2 v_i \doteq \lambda^2 \sum_{i=1}^{N} b_i^2 E(w_i - 1)^2 \tag{18.40}$$

This, in analogy to (18.29), may serve to justify points 1°, 2°, and 3° of Theorem 18.3.

The two approximations utilized in (18.40), namely replacing $\bar{\beta}$ by β and of v_i from (16.28) by $E(w_i - 1)^2$ have a tendency to compensate each other, but we shall not go into these details.

Thus some sampling-estimating strategies have the power to extend the validity of conclusions beyond the scope of strict satisfaction of Bayesian assumptions. Another failure of these assumptions, namely the violation of the type of regression, is checked by unbiasedness or approximate unbiasedness with respect to sampling.

Remark 18.5 Methods of obtaining numbers b_i from previous experience will be discussed in next chapters. We shall not go into any details concerning the costs c_i. They may depend on the distance of the unit from a center, on its size (number of subunits, for example), expected number of repeated visits, etc.

Remark 18.6 Bayesian unbiasedness expressed by $M\hat{Y} = MY$ is equivalent to representativeness defined by (16.19). In both cases (16.19) must hold for all samples that may be selected.

PROBLEMS AND NOTES

18.1 Consider an arbitrary sampling-estimating strategy (P,w). Then, under regression model (18.1),

$$\mathrm{ME}(\hat{Y} - Y)^2 \geq \lambda^2 \sum_{i=1}^{N} b_i^2 \left(\frac{1}{\pi_i} - 1\right)[1 - 2(1 - Ew_i)]$$

where $\hat{Y} = \Sigma_{i\in s}\, y_i w_i$.

18.2 Let $e = e(s,y)$ be a general (possibly nonlinear) unbiased estimate of Y. Then, under regression model (18.1),

$$ME(e - Y)^2 \geq \lambda^2 \sum_{i=1}^{N} b_i^2 \left(\frac{1}{\pi_i} - 1\right)$$

(Godambe and Joshi, 1965).

18.3 Under the regression model

$$y_i = \beta x_i + \lambda x_i e_i \qquad 1 \leq i \leq N$$

the Bayesian estimate equals

$$\hat{Y}_B = y + (X - x) \frac{1}{n} \sum_{i \in s} \frac{y_i}{x_i}$$

where n = size s.

18.4 Estimate $\hat{Y}_B$ given in (18.3) is unbiased under rejective sampling with drawing probabilities $\alpha_i = x_i/X$, $1 \leq i \leq N$.

18.5 Consider model (18.13), the Bayesian estimate $\hat{Y}_B$ given by (18.14), and the simple estimate $\hat{Y}$. Then the strategy

$$\hat{Y}_B \text{ and rejective sampling with } \alpha_i = \frac{x_i}{X} \qquad 1 \leq i \leq N \quad (18.41)$$

gives a worse expected mean square error than the strategy

$$\hat{Y} \text{ and rejective sampling with } \pi_i = \frac{nx_i}{X} \qquad 1 \leq i \leq N \quad (18.42)$$

This is true in spite of the fact that the strategy (18.41) is better on every particular sample. Explain this paradox!

18.6 Let n be fixed and consider positive vectors $(x_1,\ldots,x_N)$ such that $N \to \infty$ but $\max_{1 \leq i \leq N} x_i/\bar{X}$ and $\bar{X}$ remains bounded, and the coefficient of variation V_* satisfies $V_*^2 \geq \varepsilon > 0$, where ε is independent of N. Then, under regression model (18.13) strategy (18.42) is better than (18.41) if N is sufficiently large. Again, by "better" we mean that the expected variance $ME(\hat{Y} - Y)^2$ is smaller.

Consequently, in practice, we may expect that (18.42) is superior to (18.41) more often than it is not.

18.7 The strategy described in Theorem 18.3 also minimizes C_P given an upper bound for $\text{ME}(\hat{Y} - Y)^2$.

18.8 If the optimum π_i terms are π_i^0 and we utilize some different π_i terms, then the relative increment of $\text{ME}(\hat{Y} - Y)^2$ equals $V_*^2(1 - \varepsilon)$, where V_* is the coefficient of variation of the ratios π_i^0/π_i with the weights $c_i\pi_i$, $1 \le i \le N$, and $\varepsilon = C\,\Sigma_{i=1}^N b_i^2/(\Sigma_{i=1}^N b_i\sqrt{c_i})^2 \sim n/N$. *Hint*: Use Eq. (18.38) and Problem 1.8.

18.9 Let the optimum inclusion probabilities be

$$\pi_i^0 = \frac{n}{N}\left[1 + \alpha\left(\frac{X_i}{X} - \frac{1}{N}\right)\right]$$

and let the probabilities used be $\pi_i = n/N$, $1 \le i \le N$. Compute the relative increment of $\text{ME}(\hat{Y} - Y)^2$ if $c_i = 1$, $\alpha = 1/2$, $V_* = .5$, and $\varepsilon = .2$. [*Answer*: $\alpha^2V_*^2/(1 - \varepsilon) = (1/16)/(.8) = .078$.]

18.10 For the estimate $\hat{Y}_B$ of (18.14) and regression model (18.13) we have

$$M(\hat{Y}_B - Y)^2 = \lambda^2\left[\frac{(X - x)^2}{n} + \sum_{i \in r} x_i^2\right]$$

where $x = \Sigma_{i \in s} x_i$. This expression depends very little on the sample s of a fixed size n if n/N is small. Consequently, choosing the sample randomly causes very little loss in efficiency even under the strictly Bayesian point of view.

18.11 Under strategy (18.41) we have, asymptotically,

$$E(\hat{Y}_B - Y)^2 \sim \sum_{i=1}^{N} (y_i - \bar{\beta}x_i)^2\left(\frac{1}{\pi_i} - 1\right)$$

where

$$\bar{\beta} = \frac{\sum_{i=1}^N y_i(1 - \pi_i)}{\sum_{i=1}^N \alpha_i(1 - \pi_i)}$$

18.12 Let the unit i include x_i subunits on which the variable y assumes the value y_{ij}, $1 \le j \le x_i$, $1 \le i \le N$. If these values have a common expectation β and a common variance λ^2, and if

$$M[(y_{ij} - \beta)(y_{tk} - \beta)] = \begin{cases} \zeta\lambda^2 & \text{if } i = t \\ 0 & \text{otherwise} \end{cases}$$

then the variables $y_i = \Sigma_{j=1}^{x_i} y_{ij}$ satisfy the regression model

$$y_i = \beta x_i + \lambda[(1 - \zeta)x_i + \zeta x_i^2]^{1/2} e_i \qquad 1 \le i \le N$$

18.13 If a sampling-estimating (P,w) strategy is unbiased, then

$$\sum_{i=1}^{N} \pi_i E(w_i - 1)^2 \ge N - n$$

18.14 *Correlated residuals* Under the regression model

$$y_i = \beta x_i + \lambda x_i e_i \qquad 1 \le i \le N$$

where the residuals e_i form a stationary series with *convex* correlation function, the best strategy of size n is obtained by the simple estimate

$$\hat{Y} = \frac{X}{n} \sum_{i \in s} \frac{y_i}{x_i}$$

and systematic sampling with $\pi_i = nx_i/X$ (Hájek, 1959).

19

REGRESSION AND RATIO ESTIMATES

Representativeness may be achieved either solely by the estimation method, or solely by the sampling method, or by a combined effort of both. According to Theorem 18.3 the best strategy should employ the simple estimate, which means that all burden of achieving representativeness should be left to the sampling method. In practice, however, sampling methods ensuring representativeness may be too complex and involve extra expenses that are not reflected in Theorem 18.3. (As we remember, this theorem treats all sampling designs with fixed inclusion probabilities π_i as equally expensive.) Thus we may have good reasons to rely partially or even totally on the estimation method. As we shall show, the loss in precision incurred by achieving representativeness solely by the estimation method is equivalent to throwing away m observations, where m is the number of regressors under consideration. Consequently, if the number of regressors is small as compared with the expected sample size, it matters very little which way of acquiring representativeness we choose.

In this chapter we shall study regression estimates. These estimates are representative under any sampling and thus it makes sense to study them also under the simplest sampling of all--under Poisson sampling.

Assume that we consider m regressors $x_1, \ldots, x_m$, and that we expect that the residuals $y_i - \Sigma_{k=1}^m \bar{\beta}_k x_{ki}$ are significantly less variable than the values y_i themselves, where $\bar{\beta}_k$ are defined by

$$\sum_{i=1}^{N}\left(y_i - \sum_{k=1}^{m}\bar{\beta}_k x_{ki}\right)^2 t_i = \min_{\beta_1,\ldots,\beta_m}\left[\sum_{i=1}^{N}\left(y_i - \sum_{k=1}^{m}\beta_k x_{ki}\right)^2 t_i\right] \tag{19.1}$$

for some weights t_i. The population parameters $\bar{\beta}_k$ are not known to us, but we may replace them either by

1. Prior estimates β_k^0; or
2. Estimates $\hat{\beta}_k$ based on the sample.

If we have good prior estimates β_k^0, we may consider the estimate

$$Y^0 = \hat{Y} - \sum_{k=1}^{m}\beta_k^0(\hat{X}_k - X_k) \tag{19.2}$$

where $\hat{Y} = \Sigma_{i\in s}(y_i/\pi_i)$ and $\hat{X}_k = \Sigma_{i\in s}(x_{ki}/\pi_i)$ are the respective simple estimates and $X_k = \Sigma_{i=1}^{N} x_{ki}$ are the totals. Since β_k^0 are constants independent of the sample s, and

$$Y^0 - Y = \sum_{i=1}^{N}\left(y_i - \sum_{k=1}^{m}\beta_k^0 x_{ki}\right)\left(\frac{I_i}{\pi_i} - 1\right)$$

the variance of Y^0 may be established by means of the results for sample estimates. We must, however, replace the values y_i by residuals $y_i - \Sigma_{k=1}^{m}\beta_k^0 x_{ki}$. Thus formula (17.11) yields for *simple random sampling*:

$$\text{Var } Y^0 = \sum_{i=1}^{N}\left[y_i - \bar{Y} - \sum_{k=1}^{m}\beta_k^0(x_{ki} - \bar{X}_k)\right]^2 \frac{N}{n}\frac{N-n}{N-1} \tag{19.3}$$

Poisson sampling with $\pi_i = n/N$ provides, in view of (17.45),

$$\text{Var } Y^0 = \sum_{i=1}^{N}\left(y_i - \sum_{k=1}^{m}\beta_k^0 x_{ki}\right)^2\left(\frac{N}{n} - 1\right) \tag{19.4}$$

If the prior estimates β_k^0 are good, there will be a small difference between (19.4) and (19.3), and if we happen to have $Y = \Sigma_{k=1}^{N}\beta_k^0 X_k$, then, surprisingly enough, (19.4) will be slightly smaller than (19.3).

When sampling with varying probabilities, say

$$\pi_i = \frac{nz_i}{Z} \qquad 1 \le i \le N \tag{19.5}$$

then formula (17.44) implies for *successive sampling*

$$\bar{E}(Y^0 - Y)^2 = \sum_{i=1}^{N} \left[y_i - \frac{Y}{Z} z_i - \sum_{k=1}^{m} \beta_k^0 \left(x_{ki} - \frac{X_k}{Z} z_i \right) \right]^2 \times \left(\frac{Z}{nz_i} \right) \frac{Z - nz_i}{Z - z_i} \tag{19.6}$$

For Poisson sampling with the same probabilities we have

$$\operatorname{Var} Y^0 = \sum_{i=1}^{N} \left(y_i - \sum_{k=1}^{m} \beta_k^0 x_{ki} \right)^2 \left(\frac{Z}{nz_i} - 1 \right) \tag{19.7}$$

Example 19.1 Let x_i be measurements of the same variable as y_i, but carried out by a less precise method. Then we may put $\beta^0 = 1$, and Y^0 becomes the so-called *difference estimate*:

$$Y^0 = \hat{Y} - \hat{X} + X \tag{19.8}$$

Formula (19.3) for simple random sampling specializes as follows:

$$\operatorname{Var} Y^0 = \sum_{i=1}^{N} (y_i - \bar{Y} - x_i + \bar{X})^2 \frac{N}{n} \frac{N - n}{N - 1}$$

It is left to the reader to adapt the remaining formulas.

If we have no good prior estimates of the regression parameters, we may estimate them from the sample. Consistent estimates $\hat{\beta}_k$ of the parameters $\bar{\beta}_k$ defined by (19.1) may be obtained as least-squares estimates satisfying

$$\sum_{i \in s} \left(y_i - \sum_{k=1}^{m} \hat{\beta}_k x_{ki} \right)^2 \frac{t_i}{\pi_i} = \min_{\beta_1, \ldots, \beta_m} \left[\sum_{i \in s} \left(y_i - \sum_{k=1}^{m} \beta_k x_{ki} \right)^2 \frac{t_i}{\pi_i} \right] \tag{19.9}$$

If we compare (19.1) and (19.9), we can see that in (19.9) the population is replaced by the sample s, and, correspondingly, the weights t_i by the weights t_i/π_i. Thus coefficients in normal equations for $\hat{\beta}_k$ are simple estimates of the respective coefficients in normal equations for $\bar{\beta}_k$. From this fact consistency of $\hat{\beta}_k$ can be proved for reasonable sequences of populations and designs.

The *regression estimate* will be now denoted and defined as follows:

$$\tilde{Y} = \hat{Y} - \sum_{k=1}^{m} \hat{\beta}_k(\hat{X}_k - X_k) \tag{19.10}$$

where $\hat{Y}$ and $\hat{X}_k$ are again the simple estimates, as in (19.2).

If there is only one regressor, $\hat{\beta}$ is the solution of

$$\sum_{i\in s} (y_i - \hat{\beta}x_i)^2 \frac{t_i}{\pi_i} = \min_\beta \left[\sum_{i\in s} (y_i - \beta x_i)^2 \frac{t_i}{\pi_i}\right] \tag{19.11}$$

that is,

$$\hat{\beta} = \frac{\Sigma_{i\in s}\, y_i x_i t_i/\pi_i}{\Sigma_{i\in s}\, x_i^2 t_i/\pi_i} \tag{19.12}$$

and the regression estimate becomes

$$\tilde{Y} = \hat{Y} - \hat{\beta}(\hat{X} - X) \tag{19.13}$$

Example 19.2 (Ratio estimate) If

$$t_i = \frac{1}{x_i} \qquad 1 \le i \le N \tag{19.14}$$

(19.12) yields

$$\hat{\beta} = \frac{\hat{Y}}{\hat{X}} \tag{19.15}$$

which substituted into (19.13) gives

$$\tilde{Y} = \hat{Y} - \frac{\hat{Y}}{\hat{X}}(\hat{X} - X) = X\frac{\hat{Y}}{\hat{X}} \tag{19.16}$$

This special regression estimate will be called a *ratio estimate* and denoted by $\check{Y}$.

In the special case $x_i = 1$ and $\pi_i = n/N$, (19.16) becomes

$$\tilde{Y} = N \frac{\Sigma_{i\in s} y_i}{\Sigma_{i\in s} 1} = N\bar{y} \tag{19.17}$$

where $\bar{y}$ is the sample average. Note that $K = \Sigma_{i\in s} 1 \neq n$, unless the sampling is of fixed size.

Example 19.3 As we shall see, good weights are

$$t_i = \frac{1 - \pi_i}{\pi_i} \qquad 1 \leq i \leq N \tag{19.18}$$

which yields

$$\hat{\beta} = \frac{\Sigma_{i\in s} y_i x_i (1 - \pi_i)/\pi_i^2}{\Sigma_{i\in s} x_i^2 (1 - \pi_i)/\pi_i^2} \tag{19.19}$$

Especially, if the π_i terms are proportionate to the x_i terms, (19.19) becomes

$$\hat{\beta} = \frac{\hat{Y} - y}{\hat{X} - x} \tag{19.20}$$

where y and x are the sample totals. Usually there will be a small difference between (19.15) and (19.20), and they coincide if the π_i terms are uniformly equal to n/N. Estimate (19.20), as well as (19.15), may be utilized also if the π_i terms are not proportionate to the x_i terms. Then, however, it corresponds to the weights

$$t_i = \frac{1 - \pi_i}{x_i} \qquad 1 \leq i \leq N \tag{19.21}$$

Example 19.4 Assume that the regression of y on x is linear, not passing through the origin, i.e., $My_i = \alpha + \beta x_i$, $1 \leq i \leq N$. Thus we have two regressors 1 and x. If $\pi_i = n/N$ and $t_i = N/n$, $\hat{\alpha}$ and $\hat{\beta}$ will be defined by

$$\sum_{i\in s} (y_i - \hat{\alpha} - \hat{\beta}x_i)^2 = \min_{\alpha,\beta}\left[\sum_{i\in s} (y_i - \alpha - \beta x_i)^2\right] \tag{19.22}$$

The solution of (19.22) is well known to be

$$\hat{\alpha} = \bar{y} - \hat{\beta}\bar{x} \qquad \hat{\beta} = \frac{\Sigma_{i\in s}\, y_i(x_i - \bar{x})}{\Sigma_{i\in s}\, (x_i - \bar{x})^2} \tag{19.23}$$

where $\bar{y}$ and $\bar{x}$ are the sample averages. Consequently, denoting $K = \Sigma_{i\in s} 1$,

$$\tilde{Y} = \hat{Y} - \hat{\alpha}\left(\sum_{i\in s} \frac{N}{n} - N\right) - \hat{\beta}(\hat{X} - X)$$

$$= \frac{N}{n} y - \left(\frac{y}{K} - \hat{\beta}\frac{x}{K}\right)\left(\frac{K}{n} N - N\right) - \hat{\beta}\left(\frac{N}{n} x - X\right)$$

which simplifies to

$$\tilde{Y} = N\bar{y} - \hat{\beta}(N\bar{x} - X) \tag{19.24}$$

The regression estimate is well defined for any nonempty sample s and is representative with respect to all involved regressors for any design, which does not provide the empty sample with a positive probability.

THEOREM 19.1 For any nonempty sample s and for any regressor x_k we have

$$\tilde{X}_k = X_k \qquad 1 \le k \le m \tag{19.25}$$

Proof. If $y_i = x_{ki}$, the minimum value of $\Sigma_{i\in s}(y_i - \Sigma_{h=1}^{m} \beta_h x_{hi})^2 t_i/\pi_i$ is zero, and this is attained for $\hat{\beta}_k = 1$ and $\hat{\beta}_h = 0$, $1 \le h \ne k \le m$. Then

$$\tilde{X}_k = \tilde{Y} = \hat{Y} - \sum_{h=1}^{m} \hat{\beta}_h(\hat{X}_h - X_h) = \hat{X}_k - (\hat{X}_k - X_k) = X_k$$

Q.E.D.

Consequently the regression estimate may be used also with designs of variable size, such as Poisson sampling. Of course, we have to adjust it so as to exclude the empty sample. Obtaining the empty sample, however, will usually be an event of extremely small probability. Practically we may reject not only empty samples but also other "outliers," provided that the rejection rule does not invalidate significantly the adopted formulas.

Assuming that $\hat{\beta}_k$ are good estimates of $\bar{\beta}_k$, we may approximate $\tilde{Y}$ by the random variable

$$Y^* = \hat{Y} - \sum_{k=1}^{m} \bar{\beta}_k(\hat{X}_k - X_k) \tag{19.26}$$

Random variable Y* is not an estimate, since $\bar{\beta}_k$ depend on all y_i terms including those which are not in our sample. However, its variance may be used as a rough approximation for the mean square error of $\tilde{Y}$:

$$E(\tilde{Y} - Y)^2 \sim E(Y^* - Y)^2 = \text{Var } Y^* \tag{19.27}$$

Now the variance of Y* equals the variance of Y^0 if we replace β_k^0 by $\bar{\beta}_k$. Consequently formulas (19.3), (19.4), (19.6), and (19.7) are applicable. The resulting approximations may be considerably improved if the weights are modified so as to achieve tightness. For example, replacing in (19.7) β_k^0 by $\bar{\beta}_k$ and $[(Z/nz_i) - 1]$ by v_i, we obtain for *Poisson sampling* the following approximation:

$$\bar{E}(\tilde{Y} - Y)^2 = \sum_{i=1}^{N} \left(y_i - \sum_{k=1}^{m} \bar{\beta}_k x_{ki}\right)^2 v_i \tag{19.28}$$

where $\bar{\beta}_k$ are the solutions of (19.1). According to Theorem 16.1, formula (19.28) will be tight if the v_i terms satisfy (16.28), since (16.28) is satisfied for any weights t_i. We could also obtain a more workable approximation, which is almost tight:

$$\bar{E}(\tilde{Y} - Y)^2 = \sum_{i=1}^{N} \left(y_i - \sum_{k=1}^{m} \bar{\beta}_k x_{ki}\right)^2 \left[E(w_i - 1)^2 + \frac{m}{n}\right] \tag{19.29}$$

where w_i are the weights defining the regression estimate: $\tilde{Y} = \Sigma_{i \in s} y_i w_i$. They will be explicitly expressed later in the present chapter.

If

$$Y = \sum_{k=1}^{m} \bar{\beta}_k X_k \tag{19.30}$$

holds, formula (19.29) may be applied also to *simple random* and *successive sampling*, as is apparent from (19.3) and (19.6).

If (19.30) does not hold, (19.3) yields for simple random sampling

$$\bar{E}(\tilde{Y} - Y)^2 = \sum_{i=1}^{N} \left[y_i - \bar{Y} - \sum_{k=1}^{m} \bar{\beta}_k(x_{ki} - \bar{X}_k)\right]^2 \times \left[E(w_i - 1)^2 + \frac{m + 1}{n}\right] \tag{19.31}$$

and for successive sampling with inclusion probabilities $\pi_i = nz_i/Z$,

$$\bar{E}(\hat{Y} - Y)^2 = \sum_{i=1}^{N} \left[y_i - \frac{Y}{Z} z_i - \sum_{k=1}^{m} \bar{\beta}_k\left(x_{ki} - \frac{X_k}{Z} z_i\right)\right]^2 \times \left[E(w_i - 1)^2 + \frac{m + 1}{n}\right] \tag{19.32}$$

Since the residuals in (19.32) are smaller in the average than in (19.29), we replaced the factor m by m + 1, considering z as an additional regressor. More precise values of v_i, if so desired, could be obtained by closer analysis of (16.28) in particular cases. This is, however, not very often needed, since the factor (m + 1)/n is very small in comparison with $E(w_i - 1)^2$, and its role is rather symbolic.

The minimum value of the approximation (19.28) will be obtained for $\bar{\beta}_k$ satisfying (19.30) and (16.31), i.e., for $t_i = v_i$. Since v_i may be approximated by $(1/\pi_i) - 1$, as we shall see, the weights given by (19.18) will be nearly optimal. However, the optimum with respect to the t_i terms is very satisfactory, and there will be only a small loss of efficiency even if the weights depart considerably from the v_i terms. This is true namely when the underlying regression model fits the population well.

Example 19.5 (Ratio estimate) Consider the ratio estimate (19.16), $\check{Y} = X(\hat{Y}/\hat{X})$, which corresponds to the weights $t_i = 1/x_i$. Since $\Sigma_{i=1}^N (y_i - \beta x_i)^2 t_i$ is minimized for

$$\bar{\beta} = \frac{\sum_{i=1}^N y_i x_i t_i}{\sum_{i=1}^N x_i^2 t_i} \tag{19.33}$$

the weights $t_i = 1/x_i$ provide

$$\bar{\beta} = \frac{Y}{X} \tag{19.34}$$

Since $Y = \bar{\beta}X = (Y/X)X$ holds, (19.30) is satisfied, and we may apply (19.29) with $m = 1$:

$$\bar{E}(\check{Y} - Y)^2 = \sum_{i=1}^N \left(y_i - \frac{Y}{X} x_i\right)^2 \left[E(w_i - 1)^2 + \frac{1}{n}\right] \tag{19.35}$$

In the special case of (19.17) we have

$$\bar{E}(N\bar{y} - Y)^2 = \sum_{i=1}^N (y_i - \bar{Y})^2 \left[E(w_i - 1)^2 + \frac{1}{n}\right] \tag{19.36}$$

Example 19.6 Regression estimate $\tilde{Y} = \hat{Y} - \hat{\beta}(\hat{X} - X)$, where $\hat{\beta}$ is given by (19.20), corresponds to the weights $t_i = (1 - \pi_i)/x_i$, as observed in Example 19.3. Substituting these weights into (19.33), we obtain $\bar{\beta} = \Sigma_{i=1}^N y_i(1 - \pi_i)/\Sigma_{i=1}^N x_i(1 - \pi_i)$. Thus

$$\bar{E}(\tilde{Y} - Y)^2 = \sum_{i=1}^N (y_i - \bar{\beta}x_i)^2 \left[E(w_i - 1)^2 + \frac{1}{n}\right] \tag{19.37}$$

As in Chapter 17, we can argue that $\bar{\beta} = \Sigma_{i=1}^N y_i(1 - \pi_i)/\Sigma_{i=1}^N x_i(1 - \pi_i)$ can be replaced by Y/X if the π_i terms are small. Then (19.37) and (19.35) will coincide.

Example 19.7 Consider the situation described in Example 19.4. Then, obviously,

$$\bar{\alpha} = \bar{Y} - \bar{\beta}\bar{X} \qquad \bar{\beta} = \frac{\sum_{i=1}^N y_i(x_i - \bar{X})}{\sum_{i=1}^N (x_i - \bar{X})^2} \tag{19.38}$$

where $\bar{Y}$ and $\bar{X}$ are the population averages. Since

$$Y = \bar{\alpha}N + \bar{\beta}X = (\bar{Y} - \bar{\beta}\bar{X})N + \bar{\beta}X = Y$$

we may apply (19.29):

$$\bar{E}(\tilde{Y} - Y)^2 = \sum_{i=1}^{N} [y_i - \bar{Y} - \bar{\beta}(x_i - \bar{X})]^2 \left[E(w_i - 1)^2 + \frac{2}{n}\right] \quad (19.39)$$

where $\tilde{Y}$ is given by (19.24).

ANALYSIS OF $E(w_i - 1)^2$ Let us represent the weights w_i corresponding to a regression estimate $\tilde{Y}$ in the form

$$w_i = \frac{1}{\pi_i}(1 - \Delta_i) \qquad 1 \le i \le N \qquad i \in s \quad (19.40)$$

and denote by E_i the conditional expectation given $I_i = 1$ (i.e., $s \ni i$). Then

$$E(w_i - 1)^2 = (1 - \pi_i)1 + \pi_i E_i\left(\frac{1}{\pi_i} - 1 - \frac{\Delta_i}{\pi_i}\right)^2$$

$$= 1 - \pi_i + \pi_i\left(\frac{1}{\pi_i} - 1\right)^2 - 2\left(\frac{1}{\pi_i} - 1\right)E_i(\Delta_i) + \frac{1}{\pi_i}E_i(\Delta_i)^2$$

$$= \left(\frac{1}{\pi_i} - 1\right)\left[1 - 2E_i(\Delta_i) + \frac{E_i(\Delta_i^2)}{1 - \pi_i}\right] \quad (19.41)$$

For example, if m = 1 and $\hat{\beta}$ is given by (19.12), we have

$$\Delta_i = \frac{x_i t_i}{\sum_{i\in s} \frac{x_i^2 t_i}{\pi_i}}(\hat{X} - X) \qquad 1 \le i \le N \quad (19.42)$$

This expression is too complicated, and we have to simplify it. Noting that $\Sigma_{i\in s} x_i^2 t_i/\pi_i$ is a simple estimate for $\Sigma_{i=1}^N x_i^2 t_i$, we shall put

$$\frac{1}{\sum_{i\in s} x_i^2 t_i/\pi_i} \doteq \frac{1}{\sum_{i=1}^N x_i^2 t_i} - \frac{\sum_{i\in s} x_i^2 t_i/\pi_i - \sum_{i=1}^N x_i^2 t_i}{(\sum_{i=1}^N x_i^2 t_i)^2}$$

Consequently, Δ_i may be approximated by

$$\Delta_i^0 = x_i t_i \left(\frac{1}{U} - \frac{\hat{U} - U}{U^2}\right)(\hat{X} - X) \qquad 1 \le i \le N \tag{19.43}$$

where

$$U = \sum_{i=1}^{N} x_i^2 t_i \qquad \hat{U} = \sum_{i \in s} \frac{x_i^2 t_i}{\pi_i} \tag{19.44}$$

Now interpreting I_i/π_i as a special simple estimate corresponding to a variable z such that $z_i = 1$ and $z_j = 0$ if $j \ne i$, we may put $E_i(\hat{X} - X) = \text{Cov}(Z_i, \hat{X})$. Consequently, we have (cf. Problem 17.2)

$$\bar{E}_i(\hat{X} - X) = \begin{cases} x_i\left(\dfrac{1}{\pi_i} - 1\right) & \text{for Poisson sampling} \\ (x_i - X)\left(\dfrac{N}{n} - 1\right) & \text{for simple random sampling} \\ \left(x_i - \dfrac{X}{n}\pi_i\right)\left(\dfrac{1}{\pi_i} - 1\right) & \text{for successive sampling} \end{cases} \tag{19.45}$$

Now we shall obtain an approximation for $E(w_i - 1)^2$ by

1. Replacing Δ_i by Δ_i^0.
2. Deleting the conditioning by $I_i = 1$ in expressions

$$E_i[(\hat{U} - U)(\hat{X} - X)] \quad \text{and} \quad E_i(\hat{X} - X)^2$$

3. Ignoring terms of smaller order than those retained.

Thus we obtain

$$E_i(\Delta_i) \doteq E_i(\Delta_i^0) \doteq \frac{x_i t_i}{U} E_i(\hat{X} - X) - \frac{x_i t_i}{U^2} \text{Cov}(\hat{U}, \hat{X}) \tag{19.46}$$

and

$$E_i(\Delta_i^2) \doteq \frac{x_i^2 t_i^2}{U^2} \text{Var}(\hat{X}) \qquad 1 \le i \le N \tag{19.47}$$

Substituting (19.46) and (19.47) into (19.41), we may write

$$\bar{E}(w_i - 1)^2 = \left(\frac{1}{\pi_i} - 1\right)\left[1 - 2\frac{x_i t_i}{U} E_i(\hat{X} - X) + 2\frac{x_i t_i}{U^2}\operatorname{Cov}(\hat{U},\hat{X}) + \frac{x_i^2 t_i^2}{U^2(1 - \pi_i)} \operatorname{Var}\hat{X}\right] \qquad 1 \le i \le N \tag{19.48}$$

where U is given by (19.44) and $E_i(\hat{X} - X)$ by (19.45).

Making use of (19.45) and Problem 17.2, we obtain for *Poisson sampling*

$$\bar{E}(w_i - 1)^2 = \left(\frac{1}{\pi_i} - 1\right)\left[1 - 2\frac{x_i^2 t_i}{U}\left(\frac{1}{\pi_i} - 1\right) + 2\frac{x_i t_i}{U^2}\sum_{j=1}^{N} x_j^3 t_j\left(\frac{1}{\pi_i} - 1\right) + \frac{x_i^2 t_i^2}{U^2(1 - \pi_i)}\sum_{j=1}^{N} x_j^2\left(\frac{1}{\pi_j} - 1\right)\right] \tag{19.49}$$

Similarly for *simple random sampling* we have

$$\bar{E}(w_i - 1)^2 = \left(\frac{N}{n} - 1\right)\left[1 - 2\frac{x_i t_i (x_i - \bar{X})}{U}\left(\frac{N}{n} - 1\right) + 2\frac{x_i t_i}{U^2}\sum_{j=1}^{N} x_j^2 t_j (x_j - \bar{X})\left(\frac{N}{n} - 1\right) + \frac{x_i^2 t_i^2}{U^2}\sum_{j=1}^{N}(x_j - \bar{X})^2 \frac{N}{n}\right] \tag{19.50}$$

Finally, for *successive sampling*

$$\bar{E}(w_i - 1)^2 = \left(\frac{1}{\pi_i} - 1\right)\left\{1 - 2\frac{x_i t_i [x_i - (X/n)\pi_i]}{U}\left(\frac{1}{\pi_i} - 1\right) + 2\frac{x_i t_i}{U^2}\sum_{j=1}^{N} x_j^2 t_j\left(x_j - \frac{X}{n}\pi_j\right)\left(\frac{1}{\pi_j} - 1\right) + \frac{x_i^2 t_i^2}{U^2(1 - \pi_i)}\sum_{j=1}^{N}\left(x_j - \frac{X}{n}\pi_j\right)^2\left(\frac{1}{\pi_j} - 1\right)\right\} \tag{19.51}$$

Example 19.8 (Ratio estimate) For the ratio estimate $\check{Y} = X(\hat{Y}/\hat{X})$ the weights t_i equal $1/x_i$ so that formulas (19.49) through (19.51) may be simplified. For example, under simple random sampling, we have

$$\bar{E}(w_i - 1)^2 = \left(\frac{N}{n} - 1\right)\left[1 - \frac{2}{n}\left(1 - \frac{n}{N}\right)\frac{x_i - \bar{X}}{\bar{X}} + \frac{1}{n}\left(3 - 2\frac{n}{N}\right)V_x^2\right] \tag{19.52}$$

where V_x is the coefficient of variation of x. We can see that $\bar{E}(w_i - 1)^2$ differs from $(N/n) - 1 = (1/\pi_i) - 1$ by a disturbance which is of order $1/n$ and tends to zero if $n \to \infty$. They may be observed also in formulas (19.49) through (19.51) and justifies the approximations (19.48).

Example 19.9 Consider regression estimate (19.24) in connection with simple random sampling. Then $N\bar{y}$ and $N\bar{x}$ are simple estimates and the Δ_i terms defining the weights w_i equal

$$\Delta_i = \frac{n(x_i - \bar{x})(\bar{x} - \bar{X})}{\Sigma_{i \in s}(x_i - \bar{x})^2} \qquad 1 \le i \le N \tag{19.53}$$

Approximating Δ_i by

$$\Delta_i^0 = \frac{(x_i - \bar{X})(\bar{x} - \bar{X})}{\sigma_x^2} - \frac{(x_i - \bar{X})(\bar{x} - \bar{X})}{n\sigma_x^4}\left[\sum_{i \in s}(x_i - \bar{x})^2 - n\sigma_x^2\right]$$
$$- \frac{(\bar{x} - \bar{X})^2}{\sigma_x^2} \qquad 1 \le i \le N \tag{19.54}$$

and neglecting possible correlation between $\bar{x}$ and $\Sigma_{i \in s}(x_i - \bar{x})^2$, we may derive, similarly as above,

$$\bar{E}(w_i - 1)^2 = \left(\frac{N}{n} - 1\right)\left[1 + \frac{2}{n}\left(1 - \frac{n}{N}\right)\right.$$
$$\left. - \frac{1}{n}\left(1 - \frac{2n}{N}\right)\frac{(x_i - \bar{X})^2}{\sigma_x^2}\right] \qquad 1 \le i \le N \tag{19.55}$$

where $\sigma_x^2 = (1/N)\,\Sigma_{i=1}^N (x_i - \bar{X})^2$.

In general, multiparameter formulas for $E(w_i - 1)^2$ are tedious. However, more instructive are certain weighted averages of the excess of $E(w_i - 1)^2$ over $(1/\pi_i) - 1$:

THEOREM 19.2 Consider a regression estimate $\tilde{Y}$ corresponding to m regressors $x_1, \ldots, x_m$, and let $x_i = \Sigma_{k=1}^{m} \alpha_k x_{ki}$, $1 \le i \le N$, for some constants α_k. Then

$$\sum_{i=1}^{N} x_i \frac{E(w_i - 1)^2}{(1/\pi_i) - 1} = X + E\left[\sum_{i \in s} \frac{x_i \Delta_i^2}{\pi_i(1 - \pi_i)}\right] \tag{19.56}$$

Proof. Starting from (19.41), we have

$$\sum_{i=1}^{N} x_i \frac{E(w_i - 1)^2}{(1/\pi_i) - 1} = X - 2 \sum_{i=1}^{N} x_i E_i(\Delta_i) + \sum_{i=1}^{N} \frac{x_i}{1 - \pi_i} E_i(\Delta_i^2) \tag{19.57}$$

Now since $\tilde{Y}$ is representative relative to every regressor, it is also representative to all linear combinations of the regressors. Thus $\tilde{X} = X$, i.e.,

$$\sum_{i \in s} \frac{x_i}{\pi_i}(1 - \Delta_i) = \hat{X} - \sum_{i \in s} \frac{x_i \Delta_i}{\pi_i} = X \tag{19.58}$$

from which it follows that

$$\sum_{i=1}^{N} x_i E_i(\Delta_i) = E\left[\sum_{i \in s} \frac{x_i \Delta_i}{\pi_i}\right] = E(\hat{X} - X) = 0 \tag{19.59}$$

Now it remains to note that the last term in (19.56) equals the last term in (19.57). Q.E.D.

Relation (19.56) indicates that the success of our effort to diminish $E(w_i - 1)^2$ will depend on how we suppress the Δ_i terms, i.e., in view of (19.40), how the weights w_i for the regression estimate will be close to the weights for the simple estimate. Consequently, the other way round, we come to the same conclusion as in Theorem 18.3: a good strategy should employ an estimate which is as close to the simple estimate as possible.

In the case of several regressors, let us introduce m × m matrices

$$U = \left\{ \sum_{i=1}^{N} x_{ki} x_{hi} t_i \right\}_{k,h=1}^{m}$$
$$\hat{U} = \left\{ \sum_{i \in s} \frac{x_{ki} x_{hi} t_i}{\pi_i} \right\}_{k,h=1}^{m} \tag{19.60}$$

Then, in matrix notation,

$$\Delta_i = t_i (x_{1i}, \ldots, x_{mi}) \hat{U}^{-1} (\hat{X}_1 - X_1, \ldots, \hat{X}_m - X_m)' \qquad 1 \le i \le N \tag{19.61}$$

where $\hat{U}^{-1}$ is the inverse of $\hat{U}$. From (19.61) it is apparent that the Δ_i terms may be reduced by reducing the differences $\hat{X}_k - X_k$. In the ideal case, when

$$\hat{X}_k = X_k \qquad 1 \le k \le m \tag{19.62}$$

the Δ_i terms vanish and the regression estimate becomes identical with the simple estimate.

How much can we win if we require the sampling method to satisfy (19.62) exactly, or at least approximately? At one extreme, we may utilize Poisson sampling, under which the differences $\hat{X}_k - X_k$ are widest. Assuming that $t_i = (1 - \pi_i)/\pi_i$ and that π_i may be expressed as a linear combination of regressors, (19.56) yields

$$\sum_{i=1}^{N} \pi_i \frac{E(w_i - 1)^2}{(1/\pi_i) - 1} = n + E\left(\sum_{i \in s} \frac{\Delta_i^2}{1 - \pi_i} \right) \tag{19.63}$$

Now, in view of (19.61),

$$\begin{aligned} \sum_{i \in s} \frac{\Delta_i^2}{1 - \pi_i} &= (\hat{X}_1 - X_1, \ldots, \hat{X}_m - X_m) \hat{U}^{-1} \hat{U} \hat{U}^{-1} (\hat{X}_1 - X_1, \ldots, \hat{X}_m - X_m)' \\ &= (\hat{X}_1 - X_1, \ldots, \hat{X}_m - X_m) \hat{U}^{-1} (\hat{X}_1 - X_1, \ldots, \hat{X}_m - X_m)' \\ &\doteq (\hat{X}_1 - X_1, \ldots, \hat{X}_m - X_m) U^{-1} (\hat{X}_1 - X_1, \ldots, \hat{X}_m - X_m)' \end{aligned} \tag{19.64}$$

Furthermore, if $t_i = (1 - \pi_i)/\pi_i$, the elements of U in (19.60) are covariances of $\hat{X}_k$ and $\hat{X}_h$ under Poisson sampling (see Problem 17.2). Consequently, as is well known,

$$E[(\hat{X}_1 - X_1, \ldots, \hat{X}_m - X_m)U^{-1}(\hat{X}_1 - X_1, \ldots, \hat{X}_m - X_m)'] = m \tag{19.65}$$

Thus, making use of (19.64) and (19.65), we may simplify (19.63) as follows:

$$\sum_{i=1}^{N} \pi_i \frac{E(w_i - 1)^2}{(1/\pi_i) - 1} \doteq n + m \tag{19.66}$$

From this we may reach the following conclusion: *For any expected sample size* $n = \Sigma_{i=1}^{N} \pi_i$ *the loss incurred by utilizing Poisson sampling instead of sampling ensuring* $\hat{X}_k = X_k$, $1 \leq k \leq m$, *is approximately equivalent to throwing away* m *observations*. Obviously, the loss may be significant when the number of regressors is large, as in deeply stratified sampling. If the weights t_i are not equal to $(1 - \pi_i)/\pi_i$, the loss will be probably even greater, as is indicated by the case m = 1.

PROBLEMS AND NOTES

19.1 *Refined formula for the ratio estimate* A more precise formula for the mean square error of the ratio estimate (19.16) under simple random sampling is

$$\bar{E}(\check{Y} - Y)^2 = \sum_{i=1}^{N} \left(y_i - \frac{Y}{X} x_i\right)^2 v_i \left[1 + 6 \frac{n-1}{n^2}\left(1 - \frac{n}{N}\right)\left(\zeta^2 - \frac{1}{N}\right)V_x^2\right]$$

where ζ is the correlation coefficient of $y_i - (Y/X)x_i$ and x_i, the v_i terms are the solution of

$$v_i = E(w_i - 1)^2\left[1 - \frac{2x_i}{X} + \left(\frac{1}{X}\right)^2 \sum_{\alpha=1}^{N} \frac{x_\alpha^2 v_\alpha}{v_i}\right]^{-1} \qquad 1 \leq i \leq N$$

and V_x is the coefficient of variation of x.

For $y_i = 1$, $1 \leq i \leq N$, we have $\zeta = 1$.

19.2 *Ratio estimate and simple estimate* In large samples we have, under simple random sampling,

$$E(\check{Y} - Y)^2 \sim \left[1 - \zeta^2 + \left(\zeta - \frac{V_x}{V_y}\right)^2\right] E(\hat{Y} - Y)^2 \tag{19.67}$$

where ζ is the correlation coefficient of y and x and V_y and V_x are the respective coefficients of variation.

19.3 *Regression estimate and simple estimate* In large samples we have, under simple random sampling,

$$E(\tilde{Y} - Y)^2 \sim (1 - \zeta^2)E(\hat{Y} - Y)^2$$

where ζ is the correlation coefficient of y and x.

19.4 Using (19.67), show that asymptotically the ratio estimate is better than the simple estimate if

$$\zeta > \frac{1}{2}\frac{V_x}{V_y}$$

19.5 *Regression estimate and the Bayes estimate* Regression estimate (19.10) and the Bayes estimate (18.3) are equivalent if $b_i^2 = t_i/\pi_i$ and the inclusion probabilities π_i are a linear combination of the regressors, i.e., $\pi_i = \Sigma_{j=1}^m \lambda_j x_{ji}$, $1 \le i \le N$.

19.6 *Defined formula for the regression estimate* Put

$$\tau = \frac{\sum_{i=1}^N x_i^2(x_i - \bar{X})}{\sum_{i=1}^N (x_i - \bar{X})^2} \qquad \gamma = \frac{1}{N}\sum_{i=1}^N x_i^2 - \tau\bar{X}$$

and denote by ζ the correlation coefficient of $y_i - \bar{Y} - \bar{\beta}(x_i - \bar{X})$ and x_i^2, where $\bar{\beta} = \Sigma_{i=1}^N y_i(x_i - \bar{X})/\Sigma_{i=1}^N (x_i - \bar{X})^2$. Consider the regression estimate $\tilde{Y}$ given in (19.24).

Then, under simple random sampling, a more precise formula for $E(\tilde{Y} - Y)^2$ is

$$\bar{E}(\tilde{Y} - Y)^2 = \min_{\alpha,\beta} \sum_{i=1}^{N} (y_i - \alpha - \beta x_i)^2 v_i$$

$$\times \left[1 + \frac{4}{n-1}\left(1 - \frac{n}{N}\right)\left(\zeta^2 - \frac{1}{N}\right) \frac{N\,\Sigma_{i=1}^N (x_i^2 - \gamma - \tau x_i)^2}{[\Sigma_{i=1}^N (x_i - \bar{X})^2]^2}\right]$$

where the v_i terms satisfy the relations

$$v_i\left[1 - \frac{v_i}{\sum_{k=1}^N v_k} - \frac{v_i(x_i - \bar{X})^2}{\sum_{k=1}^N v_k(x_k - \bar{X})^2}\right] = E(w_i - 1)^2 \qquad 1 \le i \le N$$

and the weights w_i are defined by $\tilde{Y} = \Sigma_{i\in s}\, y_i w_i$.

19.7 *Bias of the ratio estimate* Consider the weights $w_i = (X/\hat{X})(I_i/\pi_i)$ corresponding to the ratio estimate. Then, under successive sampling,

$$\bar{E}w_i = 1 - \frac{(x_i/\pi_i - X/n)(1 - \pi_i)}{X} + \frac{E(\hat{X} - X)^2}{X^2} \qquad 1 \le i \le N$$

Especially, for simple random sampling, we have

$$\bar{E}w_i = 1 - \frac{1}{n}\left(1 - \frac{n}{N}\right)\left(\frac{x_i - \bar{X}}{\bar{X}} - V_x^2\right) \qquad 1 \le i \le N$$

where V_x is the coefficient of variation of x.

19.8 *Bias of the regression estimate* Let

$$w_i = \left[\frac{N}{n} - \frac{(x_i - \bar{x})(\hat{X} - X)}{\Sigma_{j\in s}(x_j - \bar{x})^2}\right] I_i$$

be the weights of the regression estimate (19.24) under simple random sampling. Then

$$\bar{E}w_i = 1 - \frac{1 - n/N}{n - 1}\left[\frac{(x_i - \bar{X})^2}{\frac{1}{N}\sum_{k=1}^N (x_k - \bar{X})^2} - 1\right] \qquad 1 \le i \le N$$

20

STRATIFICATION

As in Chapter 11, let us assume that the population is divided into L disjoint strata $S_1, \ldots, S_L$. A stratified version of a sampling method consists in L *independent* applications of the method within individual strata. Similarly, a stratified version of an estimation method will consist in its separate applications within the strata. Only for simplest methods--Poisson sampling and the simple estimate--is the stratified version equivalent to the unstratified one.

As we just mentioned, the overall simple estimate $\hat{Y}$ is a sum of simple estimates $\hat{Y}_h$ for individual strata. Actually, denoting the sample from the hth stratum by s_h, we have

$$\hat{Y} = \sum_{i \in s} \frac{y_i}{\pi_i} = \sum_{h=1}^{L} \sum_{i \in s_h} \frac{y_i}{\pi_i} = \sum_{h=1}^{L} \hat{Y}_h \tag{20.1}$$

Let us put

$$N_h = \sum_{i \in S_h} 1 \qquad Y_h = \sum_{i \in S_h} y_i \qquad X_h = \sum_{i \in S_h} x_i \quad \text{etc.} \tag{20.2}$$

and consider *stratified successive sampling* with strata sizes n_h and with

$$\pi_i = \frac{n_h x_i}{X_h} \qquad i \in S_h \qquad 1 \le h \le L \tag{20.3}$$

Then the simple estimate may be rewritten as follows:

$$\hat{Y} = \sum_{h=1}^{L} \frac{X_h}{n_h} \sum_{i \in S_h} \frac{y_i}{x_i} \tag{20.4}$$

Furthermore, since the estimates $\hat{Y}_h$ are independent under stratified sampling, we have

$$E(\hat{Y} - Y)^2 = \sum_{h=1}^{L} E(\hat{Y}_h - Y_h)^2 \tag{20.5}$$

Applying formula (17.44) to individual strata, we obtain the following approximations for $E(\hat{Y}_h - Y)^2$:

$$\bar{E}(\hat{Y}_h - Y_h)^2 = \sum_{i \in S_h} \left(y_i - \frac{Y_h}{X_h} x_i\right)^2 \frac{X_h}{n_h x_i} \frac{X_h - n_h x_i}{X_h - x_i} \tag{20.6}$$

Consequently, in view of (20.5), we may approximate $E(\hat{Y} - Y)^2$ by

$$\bar{E}(\hat{Y} - Y)^2 = \sum_{h=1}^{L} \sum_{i \in S_h} \left(y_i - \frac{Y_h}{X_h} x_i\right)^2 \frac{X_h}{n_h x_i} \frac{X_h - n_h x_i}{X_h - x_i} \tag{20.7}$$

Remembering that $h = h(i)$, we may also write, equivalently,

$$\bar{E}(\hat{Y} - Y)^2 = \sum_{i=1}^{N} \left(y_i - \frac{Y_h}{X_h} x_i\right)^2 \frac{X_h}{n_h x_i} \frac{X_h - n_h x_i}{X_h - x_i} \tag{20.8}$$

Three most important special cases of the strategy "simple estimate and stratified successive sampling" are given in Table 20.1, where $\bar{Y}_n$ denote the strata averages. Conditions on n_h for proportionality are expressed by (11.6) and (11.8).

Approximate mean square error of a *regression* estimate under stratified sampling may be derived by the method used in Chapter 19: we replace y_i by $y_i - \Sigma_{k=1}^{m} \bar{\beta}_k x_{ki}$ and modify the weights so as to achieve tightness, at least approximately. Thus, under *stratified successive sampling* with

$$\pi_i = \frac{n_h z_i}{Z_h} \qquad i \in S_h \qquad 1 \le h \le L \tag{20.9}$$

TABLE 20.1 Special cases of Eqs. (20.3), (20.4), and (20.8)

Stratified sampling	π_i	$\hat{Y}$	$\bar{E}(\hat{Y} - Y)^2$
Proportionate simple random	$\frac{n}{N}$	$\frac{N}{n} \sum_{i \in s} y_i$	$\sum_{h=1}^{L} \sum_{i \in S_h} (y_i - \bar{Y}_h)^2 \frac{N}{n}\left(1 - \frac{n}{N}\right) \frac{N_h}{N_h - 1}$
Simple random	$\frac{n_h}{N_h}$	$\sum_{h=1}^{L} \frac{N_h}{n_h} \sum_{i \in s_h} y_i$	$\sum_{h=1}^{L} \sum_{i \in S_h} (y_i - \bar{Y}_h)^2 \frac{N_h}{n_h} \frac{N_h - n_h}{N_h - 1}$
Proportionate successive	$\frac{nx_i}{X}$	$\frac{X}{n} \sum_{i \in s} \frac{y_i}{x_i}$	$\sum_{h=1}^{L} \sum_{i \in S_h} \left(y_i - \frac{Y_h}{X_h} x_i\right)^2 \frac{X}{nx_i} \frac{X_h - n_h x_i}{X_h - x_i}$

the mean square error of the *regression estimate* $\tilde{Y}$ defined by (19.10) and (19.9) may be approximated by

$$\bar{E}(\tilde{Y} - Y)^2 = \sum_{i=1}^{N} \left[y_i - \frac{Y_h}{Z_h} z_i - \sum_{k=1}^{m} \bar{\beta}_k \left(x_{ki} - \frac{X_{kh}}{Z_h} z_i\right)\right]^2 \times [E(w_i - 1)^2 + \varepsilon_i] \tag{20.10}$$

where X_{kh} is the hth stratum total of the regressor x_k, and $\bar{\beta}_k$ are given by (19.1). The weights w_i are defined by (19.46) and (19.61). Formula (20.10) is a generalization of (19.32). The adjustments ε_i are very small as compared with $E(w_i - 1)^2$. In order to have exact tightness we should put $\varepsilon_i = v_i - E(w_i - 1)^2$, where v_i is given by (16.28) applied to our case. However, it is sufficient to put

$$\varepsilon_i = \begin{cases} \dfrac{m + L - 1}{n} & \text{if (19.30) holds} \\ \dfrac{m + L}{n} & \text{otherwise} \end{cases} \tag{20.11}$$

We may also write (20.10) as follows:

$$\bar{E}(\tilde{Y} - Y)^2 = \sum_{i=1}^{N} \left(y_i - \bar{\alpha}_h z_i - \sum_{k=1}^{m} \bar{\beta}_k x_{ki}\right)^2 \times [E(w_i - 1)^2 + \varepsilon_i] \tag{20.12}$$

where the population estimates $\bar{\alpha}_h$ and $\bar{\beta}_k$ are obtained by applying least squares in *two stages*: First we minimize (19.1) and obtain $\bar{\beta}_k$; second, given $\bar{\beta}_k$ terms, we obtain $\bar{\alpha}_h$ by minimizing

$$\sum_{i=1}^{N} \left(y_i - \sum_{h=1}^{L} \alpha_h u_{hi} z_i - \sum_{k=1}^{m} \bar{\beta}_k x_{ki}\right)^2 \frac{Z_h}{n_h z_i} \tag{20.13}$$

where u_{hi} are the strata indicators:

$$u_{hi} = \begin{cases} 1 & \text{if } i \in S_h \\ 0 & \text{otherwise} \end{cases} \tag{20.14}$$

If we had wanted to obtain $\bar{\beta}_k$ from the joint minimum of

$$\sum_{i=1}^{N} \left(y_i - \sum_{h=1}^{L} \alpha_h u_{hi} z_i - \sum_{h=1}^{m} \beta_i x_{ki}\right)^2 \frac{Z_h}{n_h z_i} \tag{20.15}$$

we should have defined $\tilde{Y}$ with $\hat{\beta}_k$ obtained by minimizing

$$\sum_{i \in s} \left(y_i - \sum_{h=1}^{L} \alpha_h u_{hi} z_i - \sum_{k=1}^{m} \beta_k x_{ki}\right) \left(\frac{Z_h}{n_h z_i}\right)^2 \tag{20.16}$$

jointly in α_h and β_k.

RATIO ESTIMATE Under stratified simple random sampling and for $\tilde{Y} = \check{Y}$, where

$$\check{Y} = X \frac{\hat{Y}}{\hat{X}} = X \frac{\sum_{h=1}^{L} (N_h/n_h) \sum_{i \in s_h} y_i}{\sum_{h=1}^{L} (N_h/n_h) \sum_{i \in s_h} x_i} \tag{20.17}$$

formulas (20.12) and (20.11) yield

$$\bar{E}(\check{Y} - Y)^2 = \sum_{i=1}^{N} \left[y_i - \bar{Y}_h - \frac{Y}{X}(x_i - \bar{X}_h)\right]^2 \left[E(w_i - 1)^2 + \frac{L}{m}\right] \tag{20.18}$$

More generally, under stratified successive sampling with inclusion probabilities (20.9), we have

$$\check{Y} = X \frac{\hat{Y}}{\hat{X}} = X \frac{\sum_{h=1}^{L} (Z_h/n_h) \sum_{i \in s_h} (y_i/z_i)}{\sum_{h=1}^{L} (Z_h/n_h) \sum_{i \in s_h} (x_i/z_i)} \tag{20.19}$$

and

$$\bar{E}(\check{Y} - Y)^2 = \sum_{i=1}^{N} \left[y_i - \frac{Y_h}{Z_h} z_i - \frac{Y}{X}\left(x_i - \frac{X_h}{Z_h} z_i\right)\right]^2 \left[E(w_i - 1)^2 + \frac{L}{m}\right] \tag{20.20}$$

STRATIFIED REGRESSION ESTIMATES Applying a regression estimate $\tilde{Y}$ separately in each stratum, we obtain an estimate $\tilde{Y}_h$ for every stratum total Y_h, $1 \le h \le L$. Adding up these estimates, we obtain a stratified regression estimate $\tilde{Y}_{st}$ for Y:

$$\tilde{Y}_{st} = \sum_{h=1}^{L} \tilde{Y}_h \tag{20.21}$$

Since the regression estimates will be generally biased, with biases $B_h = E\tilde{Y}_h - Y_h$, we have

$$E(\tilde{Y}_{st} - Y)^2 = \sum_{h=1}^{L} E(\tilde{Y}_h - Y_h)^2 + \sum_{\substack{h=1 \\ h \neq k}}^{L} \sum_{k=1}^{L} B_h B_k \tag{20.22}$$

Since the biases B_h will usually all have the same sign, the bias term in (20.22) may become a considerable nuisance, namely, if the number of strata is large and if the underlying regression model does not provide a perfect fit. Then we must either modify the sampling method so as to reduce the biases or stratify the estimate less deeply.

The mean square errors $E(\tilde{Y}_h - Y_h)^2$ may be approximated by formulas derived in Chapter 19. As for the biases B_h, representing $\tilde{Y}_{st}$ in the form $\tilde{Y}_{St} = \Sigma_{i=1}^{N} y_i w_i$, we have

$$B_h = \sum_{i \in S_h} y_i (Ew_i - 1) \qquad 1 \le h \le L \tag{20.23}$$

Unbiasedness is equivalent to (18.30).

STRATIFIED RATIO ESTIMATE Under stratified simple random sampling the stratified ratio estimates will be denoted and defined as follows:

$$\check{Y}_{St} = \sum_{h=1}^{L} X_h \frac{\hat{Y}_h}{\hat{X}_h} = \sum_{h=1}^{L} X_h \frac{\Sigma_{i \in s_h} y_i}{\Sigma_{i \in s_h} x_i} \tag{20.24}$$

Utilizing in every stratum (19.35), its mean square error may be approximated by

$$\bar{E}(\check{Y}_{St} - Y)^2 = \sum_{h=1}^{L} \sum_{i \in S_h} \left(y_i - \frac{Y_h}{X_h} x_i\right)^2 \left[E(w_i - 1)^2 + \frac{1}{n_h}\right] + \sum_{\substack{h=1 \\ h \neq k}}^{L} \sum_{k=1}^{L} B_h B_k \tag{20.25}$$

Putting

$$\check{Y}_h = X_h \frac{\hat{Y}_h}{\hat{X}_h} \doteq \hat{Y}_h - \frac{\hat{Y}_h(\hat{X}_h - X_h)}{X_h} + \frac{Y_h(\hat{X}_h - X_h)^2}{X_h^2} \tag{20.26}$$

we may derive the following approximation for B_h:

$$\bar{B}_h = -\frac{\mathrm{Cov}(\hat{Y}_h, \hat{X}_h)}{X_h} + \frac{Y_h}{X_h^2} \mathrm{Var}\,\hat{X}_h = \frac{Y_h}{n_h} \frac{N_h - n_h}{N_h - 1} (V_{hx}^2 - V_{hyx}) \tag{20.27}$$

where V_{hx} is the coefficient of variation of x in the hth stratum, and V_{hyx} is the rel-variance of y and x in the same stratum (see Problem 1.6). We can see that the relative importance of $\bar{B}_h$ increases with decreasing n_h. If the regression of y on x passes through the origin in the hth stratum, we have $V_{hx}^2 = V_{hyx}$ and $\bar{B}_h = 0$, while the bias increases as the regression line deviates from the origin.

Since (20.23) holds, (20.27) is equivalent to

$$\bar{E}w_i = 1 - \frac{1}{n_h}\left(1 - \frac{n_h}{N_h}\right)\left(\frac{x_i - \bar{X}_h}{\bar{X}_h} - V_{hx}^2\right) \qquad i \in S_h \qquad 1 \le h \le L \tag{20.28}$$

Thus the strategy "$\check{Y}_{St}$ and stratified simple random sampling" overrepresents small units and underrepresents large units provided that x is a measure of size. Consequently, if we want to remove the bias, we have to increase the inclusion probabilities of large units. This goal may be accomplished by applying, in every stratum, the *ratio variant of simple random sampling* introduced in Chapter 13. As follows from Theorem 13.1, we have under stratified ratio variant of simple random sampling

$$Ew_i = \frac{N_h}{n_h} E\left(\frac{X_h}{\hat{X}_h} I_i\right) = \frac{N_h}{n_h} \frac{n_h}{N_h} = 1 \qquad i \in S_h \qquad 1 \le h \le L \quad (20.29)$$

which is equivalent to unbiasedness.

Moreover, the formula for $E(\check{Y}_h - Y_h)^2$ derived under simple random sampling is also valid for its ratio variant, their difference being reflected only by $E(w_i - 1)^2$. Consequently, under the stratified ratio variant of simple random sampling we can put

$$\bar{E}(\check{Y}_{St} - Y)^2 = \sum_{h=1}^{L} \sum_{i \in S_h} \left(y_i - \frac{Y_h}{X_h} x_i\right)^2 \left[E(w_i - 1)^2 + \frac{1}{n_h}\right] \qquad (20.30)$$

STRATIFICATION AND PRECISION We shall compare stratified and unstratified sampling both corresponding to the same inclusion probabilities π_i, $1 \le i \le N$. Stratification influences the precision by the change of second-order inclusion probabilities π_{ij} and, possibly, by the replacement of the estimates $\check{Y}$ and $\tilde{Y}$ by $\check{Y}_{St}$ and $\tilde{Y}_{St}$.

Let us start by comparing the unstratified simple random sampling with the stratified one, in connection with the simple estimate. In order to have coinciding inclusion probabilities the stratified sampling must be allocated so that $n_h/N_h = n/N$, $1 \le h \le L$. This allocation is called *proportionate*. From (17.11) and Table 20.1 we obtain

$$\text{Var } \hat{Y} = \begin{cases} \sum_{h=1}^{L} \sum_{i \in S_h} (y_i - \bar{Y})^2 \left(\frac{N}{n} - 1\right) \frac{N}{N - 1} & \text{unstratified} \\ \sum_{h=1}^{L} \sum_{i \in S_h} (y_i - \bar{Y}_h)^2 \left(\frac{N}{n} - 1\right) \frac{N_h}{N_h - 1} & \text{stratified proportionate} \end{cases} \qquad (20.31)$$

We can see that stratification increases the variance slightly if $\bar{Y}_h = \bar{Y}$, $1 \le h \le L$. Then the only difference is in the factors $N/(N - 1)$ and $N_h/(N_h - 1)$, and

$$\frac{N_h}{N_h - 1} > \frac{N}{N - 1} \qquad 1 \le h \le L \qquad (20.32)$$

However, if the strata averages are different, $\bar{Y}_h \neq \bar{Y}$, then

$$\sum_{i \in S_h} (y_i - \bar{Y}_h)^2 = \sum_{i \in S_h} [(y_i - \bar{Y})^2 - (\bar{Y}_h - \bar{Y})^2]$$

$$< \sum_{i \in S_h} (y_i - \bar{Y})^2 \qquad 1 \leq h \leq L \qquad (20.33)$$

For sufficiently large variability of the averages $\bar{Y}_h$ the inequalities (20.33) will be more influential than the inequalities (20.32) and stratification will improve the efficiency. When strata are formed at random, stratification brings no expected gain, as it should be. However, if their history or their own action make them more homogeneous than the population as a whole, then stratified sampling improves the precision. Consequently, one of the basic commandments of sampling is to *create strata as homogeneous as possible*, i.e., to have $\bar{Y}_h$ as variable as possible. In view of (1.33) the variability within strata decreases if the variability between strata (i.e., of the $\bar{Y}_h$) increases and vice versa.

A similar conclusion can be drawn for successive sampling, where we should try to achieve largest possible heterogeneity of the ratios Y_h/X_h, $1 \leq h \leq L$.

In connection with regression estimates, the precision may be increased not only by stratification of sampling but also by stratification of estimation. For example, comparing "Y_{St} and uniform stratified simple random sampling" and "$\check{Y}$ and unstratified simple random sampling," we have for the former strategy

$$\sum_{i \in S_h} \left(y_i - \frac{Y_h}{X_h} x_i\right)^2 = \sum_{i \in S_h} \left(y_i - \frac{Y}{X} x_i\right)^2$$

$$- \left(\frac{Y_h}{X_h} - \frac{Y}{X}\right)^2 \sum_{i \in S_h} x_i^2 \qquad (20.34)$$

Consequently, we have to try to create the strata so as to increase the variability of the ratios Y_h/X_h, $1 \leq h \leq L$. This variability, if significant, will be a little less exploited by "$\check{Y}$ and uniform stratified simple random sampling." Actually, we have

$$\sum_{i \in S_h} \left[y_i - \bar{Y}_h - \frac{Y}{X}(x_i - \bar{X}_h)\right]^2 = \sum_{i \in S_h} \left(y_i - \frac{Y}{X} x_i\right)^2 - \left(\frac{Y_h}{X_h} - \frac{Y}{X}\right)^2 N_h \bar{X}_h^2 \tag{20.35}$$

and $\Sigma_{i \in S_h} x_i^2 > N_h \bar{X}_h^2$. On the other hand, the drawbacks of $\check{Y}_{St}$ are larger weights $E(w_i - 1)^2$ and the danger of bias, if the ratio variant of sampling is not utilized.

Remark 20.1 If we wanted to compare nonuniform stratified simple random sampling with a corresponding nonstratified procedure, the latter procedure should be *successive* sampling with $\pi_i = n_h/N_h$, $i \in S_h$, $1 \le h \le L$.

OPTIMUM ALLOCATION Optimum inclusion probabilities derived in Theorem 18.3 are valid whether we stratify or not. If we assume that the b_i terms as well as c_i terms are constant within the strata, i.e.,

$$b_i = \bar{b}_h \qquad c_i = \bar{c}_h \qquad i \in S_h \qquad 1 \le h \le L \tag{20.36}$$

then the optimum π_i terms will also be constant within the strata, and they may be realized by simple random sampling with

$$n_h = n \frac{N_h \bar{b}_h / \sqrt{\bar{c}_h}}{\sum_{\alpha=1}^{L} N_\alpha \bar{b}_\alpha / \sqrt{\bar{c}_\alpha}} \qquad 1 \le h \le L \tag{20.37}$$

where n is the total sample size.

In the latter case optimum allocation may be also derived without reference to Theorem 18.3. For example, we can see directly that for "$\hat{Y}$ and stratified simple random sampling," the variance

$$\sum_{h=1}^{L} \sum_{i \in S_h} (y_i - \bar{Y}_h)^2 \frac{N_h}{n_h} \frac{N_h - n_h}{N_h - 1} = \sum_{h=1}^{L} \frac{N_h}{n_h} \sigma_h^2 - \sum_{h=1}^{L} \sigma_h^2$$

will be minimized for fixed $C = \Sigma_{h=1}^{L} n_h \bar{c}_h$ if

$$n_h = n \frac{N_h \sigma_h / \sqrt{\bar{c}_h}}{\sum_{\alpha=1}^{L} N_\alpha \sigma_\alpha / \sqrt{\bar{c}_\alpha}} \qquad 1 \le h \le L \tag{20.38}$$

We have denoted for this instant

$$\sigma_h^2 = \frac{1}{N_h - 1} \sum_{i \in S_h} (y_i - \bar{Y}_h)^2 \qquad 1 \le h \le L \tag{20.39}$$

Now the practical problem is that exact values of $\bar{b}_h^2$ as well as of σ_h^2 are unknown to us. On the other hand, whatever estimates we may have for σ_h^2, they may be also taken as estimates of $\bar{b}_h^2$, and vice versa. Consequently, both formulas (20.37) and (20.38) lead to coincident actions.

The relative increment of $\Sigma(N_h/n_h)\sigma_h^2$ caused by utilizing sizes n_h if the optimum sizes are n_h^0 equals squared weighted correlation coefficient of the ratios n_h^0/n_h with weights $c_h n_h$, $1 \le h \le L$, as follows from (1.21). Denoting the latter coefficient by V_*^2, the relative increment of Var $\hat{Y} = \Sigma_{h=1}^{L} (N_h/n_h)\sigma_h^2 - \Sigma_{h=1}^{L} \sigma_h^2$ will equal $V_*^2/[1 - (n/N)]$. In order to have $V_*^2 > .04$, we must have

$$V_* \ge \sqrt{.04} = 20\%$$

i.e., n_h must depart considerably from n_h^0. Small deviations from n_h^0 are unconsequential. Thus we shall abandon the proportionate allocation only if it differs significantly from the optimum one.

REGRESSION MODELS Stratified strategies may be accompanied by corresponding regression models. These models are suggested by the residuals in the formulas for the mean square error. For example, the strategy "simple estimate and stratified successive sampling with $\pi_i = n_h x_i / X_h$" corresponds to the model

$$y_i = \alpha_h x_i + \lambda b_i e_i \qquad i \in S_h \qquad 1 \le h \le L \tag{20.40}$$

Utilizing the indicators (20.14), we may rewrite (20.40) as follows:

$$y_i = \sum_{h=1}^{L} \alpha_h u_{hi} x_i + \lambda b_i e_i \qquad 1 \le h \le L \tag{20.41}$$

Possible information about the b_i terms may be partly utilized by the choice of sample sizes n_h, $1 \le h \le L$.

The strategy "ratio estimate and stratified successive sampling with $\pi_i = n_h z_i / Z_h$" corresponds to the model

$$y_i = \beta x_i + \alpha_h z_i + \lambda b_i e_i \qquad i \in S_h \qquad 1 \le h \le L \tag{20.42}$$

where the α_h satisfy the side condition

$$\sum_{h=1}^{L} \alpha_h Z_h = 0$$

The strategy "stratified ratio estimate and stratified successive sampling with $\pi_i = n_h z_i / Z_h$" corresponds to

$$y_i = \beta_h x_i + \lambda b_i e_i \qquad i \in S_h \qquad 1 \le h \le L \tag{20.43}$$

and so on.

STRATIFICATION ACROSS THE UNITS Assume that the units $i = 1, 2, \ldots, N$ are clusters of elementary units and that y_i is the sum of values of a variable associated with the elementary units. Then classifying the population of elementary units into a system of disjoint strata labeled by $a \in A$, we have

$$y_i = \Sigma_a y_{ai} \qquad x_i = \Sigma_a x_{ai} \qquad 1 \le i \le N$$

where y_{ai} is the sum of values ascertained on elementary units belonging to the unit i and simultaneously to the ath stratum. The symbol x_{ai} has a similar meaning and the sums Σ_a run over all $a \in A$.

Utilizing for every $a \in A$ a separate ratio estimate, we obtain the following stratified ratio estimate:

$$\check{Y}_A = \Sigma_a Y_a = \Sigma_a X_a \frac{\hat{Y}_a}{\hat{X}_a}$$

where $\hat{Y}_a$ and $\hat{X}_a$ denote the simple estimates based on the values y_{ai} and x_{ai}, respectively.

Putting $R_a = Y_a/X_a$, we can justify the asymptotic formula

$$\bar{E}(\check{Y}_A - Y)^2 = E[\Sigma_a (\hat{Y}_a - R_a\hat{X}_a)]^2$$

For example, under successive sampling with inclusion probabilities π_i, we have

$$\bar{E}(\check{Y}_A - Y)^2 = \sum_{i=1}^{N} (y_i - \Sigma_a R_a X_{ai})^2\left(\frac{1}{\pi_i} - 1\right)$$

A tight form would be

$$\bar{E}(\check{Y}_A - Y)^2 = \Sigma_a \Sigma_b \sum_{i=1}^{N} (y_{ai} - R_a x_{ai})(y_{bi} - R_b x_{bi}) v_{abi}$$

where the weights v_{abi}, $a \in A$, $b \in A$ are chosen so that the coefficient of $y_{ai}y_{bi}$ equals $E[(w_{ai} - 1)(w_{bi} - 1)]$, where $w_{ai} = (1/\pi_i)(X_a/\hat{X}_a)$. Approximately, we may put

$$v_{abi} = E\left[(w_{ai} - 1)(w_{bi} - 1) + \frac{1}{n}\right]$$

Example 20.1 Let the units i be apartments and the elementary units be persons. Classifying the persons into age-and-sex groups labeled by $a \in A$, y_{ai} will denote the contribution of persons living in the ith apartment and belonging to the ath age-and-sex group to the total Y. Then x_{ai} may denote the number of persons living in the ith apartment and belonging to the ath group. The stratified ratio estimate $\check{Y}_A$ will be applicable if we know the total number X_a of persons belonging to the ath age-and-sex group for every $a \in A$.

21

SUBSAMPLING

We shall continue to use symbols and concepts introduced in Chapter 12 and add some new ones, trying to keep the notation as economical as possible. The ultimate-stage units (usu) will be identified with $i = 1, 2, \ldots, N$. If they are clusters of elementary units, then y_i, x_i, etc., should be understood as *totals* of values assumed on elementary units included in the usu i, $1 \le i \le N$. For example, if y is age and usu is a household, then y_i is the sum of ages of persons included in the household; x_i may denote the number of persons (elementary units) in it, etc.

The total over the rth first-stage unit (fsu) will be denoted by Y_r, X_r, etc., $1 \le r \le M$. If the fsu are divided into strata, then the total over the αth stratum will be denoted by $Y_{(\alpha)}$, $1 \le \alpha \le H$. The fact that the rth fsu belongs to the αth stratum will be denoted by $\alpha(r) = \alpha$. In symbols we have

$$Y_r = \sum_{r(i)=r} y_i \qquad Y_{(\alpha)} = \sum_{\alpha(r)=\alpha} Y_r \tag{21.1}$$

The set of indices belonging to fsu selected in the first stage of sampling will be denoted by J, $J \subset \{1,2,\ldots,M\}$, and identified with the *first-stage sample*. The sample of usu yielded by subsampling in the rth fsu will be denoted by s_r, $1 \le r \le M$. Consequently, the total sample of usu equals

$$s = \bigcup_{r \in J} s_r \tag{21.2}$$

With these symbols $\pi_r^{I} = P(r \in J)$ and $\pi_i^{II} = P(i \in s \mid r(i) \in J)$.

SIMPLE ESTIMATE The simple estimate $\hat{Y}$ is defined by (17.1), as always. The only new circumstance is the possibility of decomposing π_i according to (12.1). Consequently, we may write

$$\hat{Y} = \sum_{i \in s} \frac{y_i}{\pi_i} = \sum_{i \in s} \frac{y_i}{\pi_r^I \pi_i^{II}} = \sum_{r \in J} \frac{\sum_{i \in s_r} \frac{y_i}{\pi_i^{II}}}{\pi_r^I} = \sum_{r \in J} \frac{\hat{Y}_r}{\pi_r^I} \tag{21.3}$$

where $\hat{Y}_r$ is the simple estimate for Y_r based on subsampling:

$$\hat{Y}_r = \sum_{i \in s_r} \frac{y_i}{\pi_i^{II}} \qquad 1 \le r \le M \tag{21.4}$$

THEOREM 21.1 In multistage sampling we have

$$E(\hat{Y} - Y)^2 = E_I(\hat{Y}_I - Y)^2 + \sum_{r=1}^{M} \frac{E_{II}(\hat{Y}_r - Y_r)^2}{\pi_r^I} \tag{21.5}$$

where

$$\hat{Y}_I = \sum_{r \in J} \frac{Y_r}{\pi_r^I} \tag{21.6}$$

and the expectations $E_I(\cdot)$ and $E_{II}(\cdot)$ refer to the first-stage sampling and subsampling, respectively.

Proof. Fixing the first-stage sample J, we have

$$E_{II}\hat{Y} = \sum_{r \in J} \frac{E_{II}(\hat{Y}_r)}{\pi_r^I} = \sum_{r \in J} \frac{Y_r}{\pi_r^I} = \hat{Y}_I \tag{21.7}$$

Furthermore, since subsampling is carried out in each fsu independently, we have

$$\begin{aligned} E_{II}(\hat{Y} - Y)^2 &= (E_{II}\hat{Y} - Y)^2 + E_{II}(\hat{Y} - E_{II}\hat{Y})^2 \\ &= (\hat{Y}_I - Y)^2 + E_{II}\left[\frac{\sum_{r \in J} (\hat{Y}_r - Y_r)}{\pi_r^I}\right]^2 \end{aligned}$$

$$= (\hat{Y}_I - Y)^2 + \sum_{r \in J} \frac{E_{II}(\hat{Y}_r - Y_r)^2}{(\pi_r^I)^2} \tag{21.8}$$

Applying the well-known relation $E(\cdot) = E_I[E_{II}(\cdot)]$ to both sides of (21.8), we obtain (21.5). Q.E.D.

TERMINOLOGY We shall call $E_I(\hat{Y}_I - Y)^2$ the first-stage component of $E(\hat{Y} - Y)^2$, and $\Sigma_{r=1}^M E_{II}(\hat{Y}_r - Y_r)^2/\pi_r^I$ the subsampling component of $E(\hat{Y} - Y)^2$. If there are only two stages, the subsampling component may be also called the second-stage component.

Example 21.1 Consider two-stage sampling, the first stage of which is *successive* with

$$\pi_r^I = \frac{mZ_r}{Z} \qquad 1 \le r \le M \tag{21.9}$$

and second stages are *simple random* of sizes n_r. Let the sizes n_r be adjusted so that the overall sampling is uniform of expected size n:

$$\pi_i = \frac{mZ_r}{Z}\frac{n_r}{N_r} = \frac{n}{N} \qquad 1 \le i \le N \tag{21.10}$$

Then the simple estimate $\hat{Y}$ is given by

$$\hat{Y} = \frac{N}{n}\sum_{i \in s} y_i \tag{21.11}$$

Applying (17.44) to $E_I(\hat{Y}_I - Y)^2$ and (17.11) to $E_{II}(\hat{Y}_r - Y_r)^2$, we obtain

$$\bar{E}(\hat{Y} - Y)^2 = \frac{Z}{m}\sum_{r=1}^M \left(\frac{Y_r}{Z_r} - \frac{Y}{Z}\right)^2 Z_r \frac{Z - mZ_r}{Z - Z_r}$$
$$+ \frac{N}{n}\sum_{i=1}^N (y_i - \bar{Y}_r)^2 \frac{N_r - n_r}{N_r - 1} \qquad r = r(i) \tag{21.12}$$

where $\bar{Y}_r = Y_r/N_r$.

If $Z_r = 1$, the successive sampling turns into simple random sampling and (21.12) becomes

$$\bar{E}(\hat{Y} - Y)^2 = \frac{M}{m} \sum_{r=1}^{M} \left(Y_r - \frac{Y}{M}\right)^2 \frac{M - m}{M - 1}$$
$$+ \frac{N}{n} \sum_{i=1}^{N} (y_i - \bar{Y}_r) \frac{N_r - n_r}{N_r - 1} \tag{21.13}$$

In order to minimize the first-stage component it is appropriate to put $Z_r = N_r$, which yields

$$\bar{E}(\hat{Y} - Y)^2 = \frac{N}{m} \sum_{r=1}^{M} (\bar{Y}_r - \bar{Y})^2 N_r \frac{N - mN_r}{N - N_r}$$
$$+ \frac{N}{n} \sum_{i=1}^{N} (y_i - \bar{Y}_r)^2 \frac{N_r - \bar{n}}{N_r - 1} \tag{21.14}$$

where $\bar{n}$ is the *constant* value of n_r, $1 \le r \le M$.

Example 21.2 More generally we may assume that the successive sampling as well as simple random sampling are stratified and uniformly (proportionately) allocated. That means that the sizes m_α in the first-stage strata satisfy

$$\frac{m}{Z_{(\alpha)}} = \frac{m}{Z} \qquad 1 \le \alpha \le H \tag{21.15}$$

and the sizes n_{rh} in the second-stage strata satisfy

$$\frac{n_{rh}}{N_{rh}} = \frac{n_r}{N_r} \qquad 1 \le r \le M \qquad 1 \le h \le L_r \tag{21.16}$$

where N_{rh} is the size of the hth second-stage stratum within the rth fsu.

Then formula (21.12) remains valid and

$$\bar{E}(\hat{Y} - Y)^2 = \frac{M}{m} \sum_{r=1}^{M} \left[\frac{Y_r}{Z_r} - \frac{Y_{(\alpha)}}{Z_{(\alpha)}}\right]^2 Z_r \frac{Z_{(\alpha)} - mZ_r}{Z_{(\alpha)} - Z_r} + \frac{N}{n} \sum_{i=1}^{N} (y_i - \bar{Y}_{rh})^2$$
$$\times \frac{N_{rh} - n_{rh}}{N_{rh} - 1} \qquad \alpha = \alpha(r) \qquad rh = rh(i) \qquad (21.17)$$

where $\bar{Y}_{rh}$ is the average in the rhth secondary stratum.

THREE AND MORE STAGES If the sampling has three stages, then subsampling has two stages and we may utilize previous results to derive $E_{II}(\hat{Y}_r - Y_r)^2$, which substituted into (21.5) yields $E(\hat{Y} - Y)^2$. The formula for three-stage sampling may then be utilized to derive $E_{II}(\hat{Y}_r - Y_r)^2$ for four-stage sampling, and so on.

RATIO ESTIMATE Consider Example 21.1 and assume that for some reason we are not able to put $Z_r = N_r$. Then the variability of N_r will influence the first-stage component unfavorably and we should utilize the ratio estimate

$$\check{Y} = N \frac{\hat{Y}}{\hat{K}} = N\bar{y} = N \frac{y}{K} \qquad (21.18)$$

where $\bar{y}$ is the sample average and K is the actual sample size. In view of Chapter 19 we may approximate $E(\check{Y} - Y)^2$ by $E(\hat{Y} - \bar{Y}\hat{N})^2$ where $\bar{Y} = Y/N$, which is equivalent to replacing y_i by $y_i - \bar{Y}$ in (21.12). Thus we obtain

$$\bar{E}(\check{Y} - Y)^2 = \frac{Z}{m} \sum_{r=1}^{M} (\bar{Y}_r - \bar{Y})^2 \frac{N_r^2}{Z_r} \frac{Z - mZ_r}{Z - Z_r}$$
$$+ \frac{N}{n} \sum_{i=1}^{N} (y_i - \bar{Y}_r)^2 \frac{N_r - n_r}{N_r - 1} \qquad (21.19)$$

Example 21.3 Let the fsu be towns and N_r denote the number of apartments at the time of sampling, where Z_r denotes the number of apartments at the last census. Then making use of $\pi_r^I = mZ_r/Z$ avoids the necessity of ascertaining N_r in the towns not selected in the first stage. Of course, in order to compute $\check{Y} = N\bar{y}$ we must know N from some external sources. Alternatively, we may estimate N by

$$N_0 = \frac{Z}{m} \sum_{r \in J} \frac{N_r}{Z_r} \tag{21.20}$$

and put $\check{Y}_0 = N_0\bar{y}$. If N_0 is precise enough, we may assume that $E(\check{Y}_0 - Y)^2 = E(\check{Y} - Y)^2$. The knowledge of N is not necessary for estimating Y/X by y/x. Replacing y_i by $y_i - (Y/X)x_i$ in (21.12) and dividing by X^2, we obtain

$$\bar{E}\left(\frac{y}{x} - \frac{Y}{X}\right)^2 = \frac{Z}{mX^2} \sum_{r=1}^{M} \left(\frac{Y_r}{X_r} - \frac{Y}{X}\right)^2 \frac{X_r^2}{Z_r} \frac{Z - mZ_r}{Z - Z_r}$$

$$+ \frac{N}{n} \sum_{i=1}^{N} \left(y_i - \frac{Y}{X} x_i - \bar{Y}_r + \frac{Y}{X} \bar{X}_r\right)^2 \frac{N_r - n_r}{N_r - 1} \tag{21.21}$$

TIGHTNESS The concept of tightness as defined in Chapter 16 was appropriate for one-stage sampling designs only. It has to be modified if there are many stages. For example, if there are two stages, we shall require that the exact formula y'Ay and its approximation $y'\bar{A}y$ coincide not only on the diagonal, i.e., $\bar{a}_{ii} = a_{ii}$, but also for all pairs (i,j) belonging to the same fsu:

$$\bar{a}_{ij} = a_{ij} \qquad r(i) = r(j) \tag{21.22}$$

Formula (21.19) can be made tight in this sense if we put

$$\bar{E}(\check{Y} - Y)^2 = \sum_{r=1}^{M} (Y_r - \bar{Y}N_r)^2 v_r^I + \sum_{i=1}^{N} (y_i - \bar{Y}_r)^2 v_i^{II} \tag{21.23}$$

where the weights v_r^I satisfy the relations

$$v_r^I = E(w_r^I - 1)^2 \left(1 - \frac{2N_r}{N} + \frac{\sum_{\alpha=1}^{M} N_\alpha^2 v_\alpha^I / v_r^I}{N^2}\right)^{-1} \tag{21.24}$$

and

$$v_i^{II} = E(w_r^I)^2 \frac{N_r}{n_r} \frac{N_r - n_r}{N_r - 1} \qquad r = r(i) \tag{21.25}$$

where

$$w_r^I = \frac{n}{K} \frac{Z}{mZ_k} \qquad r \in J \qquad 1 \le r \le M \tag{21.26}$$

The weights w_r^I may be called the first-stage weights, since

$$\check{Y} = \sum_{r \in J} \hat{Y}_r w_r^I \tag{21.27}$$

In particular, setting $Z_r = 1$, we obtain

$$w_r^I = \frac{m}{K} \frac{M}{m} \qquad r \in J \qquad 1 \le r \le M \tag{21.28}$$

SYSTEMATIC SUBSAMPLING As we shall see in the next chapter, it is not essential to know the subsampling component of the mean square error. Therefore, on the second and further stages we may utilize designs for which the error cannot be robustly estimated. First of all, we can make use of the systematic sampling defined in Chapter 10. We do not even need to make it randomized.

REGRESSION MODEL In connection with two-stage sampling it is natural to assume that the residuals belonging to the same fsu are correlated. For example, we may consider the regression model

$$y_i = \alpha + \lambda^2 e_i \qquad 1 \le i \le N \tag{21.29}$$

where

$$Me_i e_j = \begin{cases} 1 & \text{if } i = j \\ \zeta & \text{if } r(i) = r(j) \\ 0 & \text{otherwise} \end{cases} \tag{21.30}$$

OPTIMUM DESIGN Consider the two-stage sampling described in Example 21.1 and the ratio estimate $\check{Y} = N\bar{y}$, and compute the expected mean square error $ME(\check{Y} - Y)^2$. Interchanging the expectations, we have under model (21.29)-(21.30)

$$M(N\bar{y} - Y)^2 = \frac{N^2}{K^2} \sum_{r \in J} n_r(1 - \zeta + \zeta n_r) - 2\frac{N}{K} \sum_{r \in J} n_r(1 - \zeta + \zeta N_r) + \sum_{r=1}^{M} N_r(1 - \zeta + \zeta N_r) \qquad (21.31)$$

where K is the actual sample size. Replacing K by its expectation n and taking then the mean value with respect to sampling, we obtain, in view of (21.10),

$$\begin{aligned} EM(N\bar{y} - Y)^2 &\doteq \frac{N^2}{n^2} \sum_{r=1}^{M} \frac{n_r m Z_r}{Z} (1 - \zeta + \zeta n_r) \\ &\quad - 2\frac{N}{n} \sum_{r=1}^{M} \frac{n_r m Z_r}{Z} (1 - \zeta + \zeta N_r) \\ &\quad + \sum_{r=1}^{M} N_r(1 - \zeta + \zeta N_r) \\ &= \zeta \sum_{r=1}^{M} N_r^2 \frac{Z}{mZ_r} + \frac{N^2}{n}(1 - \zeta) \\ &\quad - \sum_{r=1}^{M} N_r(1 - \zeta + \zeta N_r) \end{aligned} \qquad (21.32)$$

If the variable part of the costs may be expressed as follows:

$$C = \sum_{r=1}^{M} c_r^I \frac{mZ_r}{Z} + nc^{II} \qquad (21.33)$$

then for given costs $EM(\check{Y} - Y)^2$ will be minimized for

$$\frac{N_r}{N} \frac{Z}{Z_r} \frac{n}{m} = \sqrt{\frac{c_r^I}{c^{II}} \frac{1 - \zeta}{\zeta}} \qquad (21.34)$$

This may be shown by means of the Cauchy inequality (1.20) applied already in the proof of Theorem 18.3. If the costs c_r^I are constant, $c_r^I = c^I$, then (21.34) is solved for

$$Z_r = N_r \qquad \text{and} \qquad \bar{n} = \sqrt{\frac{c^I}{c^{II}} \frac{1 - \zeta}{\zeta}} \tag{21.35}$$

where $\bar{n} = n/m$ is the size of subsampling in the fsu.

Thus the optimum design is given by a sampling selecting the fsu with probabilities proportionate to the size N_r, and taking $\bar{n}$ ssu in the second stage from each selected fsu. The number $\bar{n}$ decreases as the correlation coefficient ζ within the fsu increases. Moreover, $\bar{n}$ is an increasing function of the costs c^I associated with the inclusion of one fsu and a decreasing function of the costs c^{II} associated with the inclusion of one ssu.

If the fsu are stratified into strata according to size N_r, the optimum $\bar{n}_\alpha$ in the αth stratum will equal

$$\bar{n}_\alpha = \sqrt{\frac{c_\alpha^I}{c_\alpha^{II}} \frac{1 - \zeta_\alpha}{\zeta_\alpha}} \qquad 1 \le \alpha \le H \tag{21.36}$$

Usually the optimum $\bar{n}_\alpha$ will be an increasing function of the average stratum size $\bar{N}_\alpha$, but $\bar{n}_\alpha/\bar{N}_\alpha$ will be a decreasing function of $\bar{N}_\alpha$.

22

A GENERAL APPROACH

Let $\hat{Y}$ be a linear estimate and $E(\hat{Y} - Y)^2$ its mean square error. An estimate of $E(\hat{Y} - Y)^2$ will be denoted by $\hat{E}(\hat{Y} - Y)^2$ or briefly by v. Assume that the strategy is representative with respect to $x_1, \ldots, x_m$ and that it is robust. This means that a good approximation for $E(\hat{Y} - Y)^2$ has the form (16.27).

If the regression coefficients $\bar{\beta}_1, \ldots, \bar{\beta}_m$ were known, we could construct $\hat{E}(\hat{Y} - Y)^2$ as a simple estimate:

$$\hat{E}(\hat{Y} - Y)^2 = \sum_{i \in s} \left(y_i - \sum_{k=1}^{m} \bar{\beta}_k x_{ki}\right)^2 \frac{v_i}{\pi_i} \tag{22.1}$$

If the regression coefficients are not known, as is usually the case, we can estimate them from the sample.

Assuming that $\bar{\beta}_k$ are defined as those minimizing (16.6), we may define their estimates, say $\hat{\beta}_k$, as solutions of

$$\sum_{i \in s} \left(y_i - \sum_{k=1}^{m} \beta_k x_{ki}\right)^2 \frac{t_i}{\pi_i} = \text{minimum} \tag{22.2}$$

In contrast to (16.6), the sum in (22.2) runs over the sample only, but the weights t_i are replaced by t_i/π_i. Thus the sum (16.6) is the mean value of the sum in (22.2):

$$E\left[\sum_{i \in s} \left(y_i - \sum_{k=1}^{m} \beta_k x_{ki}\right)^2 \frac{t_i}{\pi_i}\right] = \sum_{i=1}^{N} \left(y_i - \sum_{k=1}^{m} \beta_k x_{ki}\right)^2 t_i$$

Intuitively, we may expect that solutions of (22.2) will be good estimates of $\bar{\beta}_k$, which are solutions of

$$\sum_{i=1}^{N} \left(y_i - \sum_{k=1}^{m} \beta_k x_{ki}\right)^2 t_i = \text{minimum} \tag{22.3}$$

Example 22.1 Let $m = 1$, $x_i = 1$, $\pi_i = n/N$, and $t_i = 1$, $1 \le i \le N$. Then (22.3) becomes

$$\sum_{i=1}^{N} (y_i - \beta)^2 = \text{minimum}$$

which is solved by $\bar{\beta} = \bar{Y}$, the population average. Similarly, (22.2) becomes

$$\frac{N}{n} \sum_{i \in s} (y_i - \beta)^2 = \text{minimum}$$

which is solved by $\hat{\beta} = \bar{y}$, the sample average.

Example 22.2 Let $m = 1$, $\pi_i = nx_i/X$, and $t_i = 1/x_i$, $1 \le i \le N$. Then (22.3) becomes

$$\sum_{i=1}^{N} (y_i - \beta x_i)^2 \frac{1}{x_i} = \text{minimum} \tag{22.4}$$

which is solved by $\bar{\beta} = Y/X$, the population ratio. Similarly, (22.2) becomes

$$\frac{X}{n} \sum_{i \in s} (y_i - \beta x_i)^2 \frac{1}{x_i^2} = \text{minimum}$$

which is solved by $\hat{\beta} = \frac{1}{n} \Sigma_{i \in s} (y_i/x_i)$.

Example 22.3 If $m = 1$, $t_i = 1/x_i$, and π_i is arbitrary, $1 \le i \le N$, then $\bar{\beta} = Y/X$ and $\hat{\beta} = \hat{Y}/\hat{X}$, where $\hat{Y}$ and $\hat{X}$ are simple estimates.

Substituting $\hat{\beta}_k$ for $\bar{\beta}_k$ into (22.1), we obtain

$$\hat{E}(\hat{Y} - Y)^2 = \sum_{i \in s} \left(y_i - \sum_{k=1}^{m} \hat{\beta}_k x_{ki}\right)^2 \frac{v_i}{\pi_i} \tag{22.5}$$

Formula (22.5), however, will tend to underestimate $E(\hat{Y} - Y)^2$ because $\hat{\beta}_k$ are solutions of (22.2) and v_i/π_i will usually differ only slightly from $\lambda t_i/\pi_i$ for proper λ.

In order to check the tendency to underestimate in (22.5), we have to replace v_i/π_i by $\hat{v}_i/\pi_i$, where the $\hat{v}_i$ terms will be determined by an analog of tightness: We shall require that the coefficient of y_i^2 in

$$\hat{E}(\hat{Y} - Y)^2 = \sum_{i \in s} \left(y_i - \sum_{k=1}^{m} \hat{\beta}_k x_{ki}\right)^2 \frac{v_i}{\pi_i} \tag{22.6}$$

will be a good estimate of $E(w_i - 1)^2$, which is the coefficient of y_i^2 in $E(\hat{Y} - Y)^2$, as we know.

A simple and good estimate for $E(w_i - 1)^2$ may be obtained as follows. Assuming that the strategy is not too biased, we may put

$$Ew_i \doteq 1$$

and, consequently,

$$E(w_i - 1)^2 \doteq E[w_i(w_i - 1)]$$

Now, however, $E[w_i(w_i - 1)]$ may be estimated simply by $w_i(w_i - 1)$! Note that $E(w_i - 1)^2$ *cannot* be estimated by $(w_i - 1)^2$, since $(w_i - 1)^2 \neq 0$ for $i \notin s$, while the coefficient of y_i^2 in (22.6) equals zero if $i \notin s$.

DEFINITION We shall say that formula (22.6) is *tight* if the coefficient of y_i^2 equals $w_i(w_i - 1)$.

THEOREM 22.1 Let $\hat{\beta}_k$ be the solutions of (22.2) for some positive weights t_i, $1 \leq k \leq m$, $1 \leq i \leq N$. Then formula (22.6) will be tight if the $\hat{v}_i$ are solutions of

$$\frac{\hat{v}_i}{\pi_i} = w_i(w_i - 1)\left[1 - 2\sum_{k=1}^{m} \hat{a}_{ki} x_{ki} + \sum_{k=1}^{m}\sum_{h=1}^{m} \hat{a}_{ki}\hat{a}_{hi} \sum_{\alpha \in s} \frac{x_{k\alpha} x_{h\alpha} \hat{v}_\alpha}{\hat{v}_i}\right]^{-1} \qquad 1 \leq i \leq N \tag{22.7}$$

where the constants $\hat{a}_{ki}$, $1 \leq k \leq m$, $i \in s$, are defined by

$$\hat{\beta}_k = \sum_{i \in s} \hat{a}_{ki} y_i \qquad 1 \le k \le m \tag{22.8}$$

Proof. The proof is straightforward and analogous to the proof of Theorem 16.1. Q.E.D.

If m = 1, (22.6) becomes

$$\hat{E}(\hat{Y} - Y)^2 = \sum_{i \in s} \frac{(y_i - \hat{\beta} x_i)^2 \hat{v}_i}{\pi_i} \tag{22.9}$$

where

$$\hat{\beta} = \frac{\Sigma_{i \in s} y_i x_i t_i / \pi_i}{\Sigma_{i \in s} x_i^2 t_i / \pi_i} \tag{22.10}$$

and

$$\frac{\hat{v}_i}{\pi_i} = w_i(w_i - 1)\left[1 - 2\frac{x_i^2 t_i / \pi_i}{\Sigma_{\alpha \in s} x_\alpha^2 t_\alpha / \pi_\alpha} + \frac{x_i^2 t_i^2 / (\hat{v}_i \pi_i)}{(\Sigma_{\alpha \in s} x_\alpha^2 t_\alpha / \pi_\alpha)^2} \sum_{\alpha \in s} \frac{x_\alpha^2 \hat{v}_\alpha}{\pi_\alpha}\right]^{-1} \qquad 1 \le i \le N \tag{22.11}$$

Analogously to Chapter 16, the harmonic mean of the factor $[\cdot]^{-1}$ in (22.11) equals either n/(n - 1) or is somewhat less, n being the sample size. If there are m regressors, it will be n/(n - m) or slightly less.

Approximate solutions of (22.7) and (22.11) are equal:

$$\frac{\hat{v}_i}{\pi_i} = w_i(w_i - 1)\left[1 - \sum_{k=1}^{m} \hat{a}_{ki} x_{ki}\right]^{-1} \qquad 1 \le i \le N \tag{22.12}$$

and

$$\frac{\hat{v}_i}{\pi_i} = w_i(w_i - 1)\left[1 - \frac{x_i^2 t_i / \pi_i}{\Sigma_{\alpha \in s} x_\alpha^2 t_\alpha / \pi_\alpha}\right]^{-1} \tag{22.13}$$

They are satisfactory for all practical purposes and tend to overestimate $\hat{v}_i / \pi_i$ very slightly.

Substituting (22.12) into (22.6) we have, in view of (22.8),

$$\hat{E}(\hat{Y} - Y)^2 = \sum_{i \in s} \left(y_i - \sum_{k=1}^{m} \hat{\beta}_k x_{ki} \right)^2 w_i (w_i - 1)(1 - \hat{\varepsilon}_i)^{-1} \tag{22.14}$$

where

$$\hat{\varepsilon}_i = \sum_{k=1}^{m} \hat{\beta}_k x_{ki} \qquad \text{with } y_i = 1,\ y_j = 0,\ j \neq i \tag{22.15}$$

Remark 22.1 As we have said in Chapter 4, our aim is not to estimate $E(\hat{Y} - Y)^2$ but rather the error $(\hat{Y} - Y)^2$ itself. A perfect satisfaction of this goal is possible in the Bayesian theory, where we utilize all information about the expected error provided by the sample *actually obtained*. In the robust theory--as a next good thing to do--we may try to prefer error estimates which are close to Bayesian estimates.

Assume that model (18.1) holds and recall the fact that representativeness is equivalent to $M\hat{Y} = Y$. Then for a linear estimate $\hat{Y}$ and a fixed sample s provided by a representative theory, we have

$$M(\hat{Y} - Y)^2 = M\left[\sum_{i=1}^{N} y_i (w_i - 1) \right]^2 = \lambda^2 \sum_{i=1}^{N} b_i^2 (w_i - 1)^2$$

Since $w_i = 0$ for $i \notin s$, we may further write

$$M(\hat{Y} - Y)^2 = \lambda^2 \sum_{i \in s} b_i^2 w_i (w_i - 1) - \lambda^2 \sum_{i \in s} b_i^2 w_i + \lambda^2 \sum_{i=1}^{N} b_i^2 \tag{22.16}$$

As $\Sigma_{i \in s} b_i^2 w_i$ is an estimate of $\Sigma_{i=1}^{N} b_i^2$, and both terms are small in comparison with $\Sigma_{i \in s} b_i^2 w_i (w_i - 1)$, we may omit the last two terms in (22.16) without much harm. Thus we have, approximately,

$$M(\hat{Y} - Y)^2 \doteq \lambda^2 \sum_{i \in s} b_i^2 w_i (w_i - 1) \tag{22.17}$$

If the constants are known, we could estimate λ^2 by (18.9) and have the following error estimate:

$$M(\hat{Y} - Y)^2 = \hat{\lambda}^2 \sum_{i \in s} b_i^2 w_i (w_i - 1) \tag{22.18}$$

However, if our knowledge of the b_i terms is not reliable, we may prefer a more robust estimation method consisting of putting an estimate of $(\lambda^2 b_i^2)$ equal to

$$\left(y_i - \sum_{k=1}^{m} \hat{\beta}_k x_{ki}\right)^2 (1 - \hat{\varepsilon}_i)^{-1}$$

where $(1 - \hat{\varepsilon}_i)^{-1}$ is a factor balancing the effect of the substitution of $\hat{\beta}_k$ for unknown regression coefficients β_k, $1 \le k \le m$. In this way we obtain the following estimate of $M(\hat{Y} - Y)^2$:

$$\hat{M}(\hat{Y} - Y)^2 = \sum_{i \in s} \left(y_i - \sum_{k=1}^{m} \hat{\beta}_k x_{ki}\right)^2 w_i (w_i - 1)(1 - \hat{\varepsilon}_i)^{-1} \tag{22.19}$$

Comparing (22.19) with (22.14), we can see that these two error estimates will coincide if $\hat{\varepsilon}_i$ are given by (22.15).

Consequently, our error estimate (22.14) will be close to the Bayes error estimate and may be expected to be more correlated with the actual error $(\hat{Y} - Y)^2$ than, for example, an estimate of the form

$$\sum_{i \in s} \left(y_i - \sum_{k=1}^{m} \hat{\beta}_k x_{ki}\right)^2 \frac{E(w_i - 1)^2}{\pi_i} (1 - \hat{\varepsilon}_i)^{-1} \tag{22.20}$$

23

APPLICATIONS TO SIMPLE RANDOM SAMPLING

In the present chapter we shall consider several sampling-estimating strategies in which sampling is simple random, stratified or not.

SIMPLE ESTIMATE AND SIMPLE RANDOM SAMPLING For this strategy $\pi_i = n/N$, $x_i = 1$, $t_i = 1$, and $w_i = N/n$ if $i \in s$. The variance of $\hat{Y}$ is given by (17.11), and the sample regression coefficient $\hat{\beta}$ equals $\bar{y}$--the sample average, as we have seen in Example 22.1. The solutions of (22.7) will be independent of i, say $\hat{v}/\pi$, and it will equal

$$\frac{\hat{v}}{\pi} = \frac{N}{n}\left(\frac{N}{n} - 1\right)\left[1 - 2\,\frac{1}{n} + \frac{1}{n}\right]^{-1} = \frac{N(N - n)}{n(n - 1)} \tag{23.1}$$

Substituting $\hat{\beta}$ and $\hat{v}_i/\pi_i$ into (22.9) we obtain

$$\hat{E}(\hat{Y} - Y)^2 = \frac{N(N - n)}{n(n - 1)} \sum_{i \in s} (y_i - \bar{y})^2 \tag{23.2}$$

The same result would be obtained from (22.14).

THEOREM 23.1 Variance estimate (23.2) is unbiased.

Proof. Utilizing (1.32) and (5.5) and introducing the indicators of inclusion I_i, we have

$$E\left[\sum_{i\in s} (y_i - \bar{y})^2\right] = E\left[\frac{1}{2n} \sum_{i=1}^{N} \sum_{\substack{j=1 \\ i\neq j}}^{N} (y_i - y_j)^2 I_i I_j\right]$$

$$= \frac{1}{2n} \sum_{i=1}^{N} \sum_{\substack{j=1 \\ i\neq j}}^{N} (y_i - y_j)^2 E(I_i I_j)$$

$$= \frac{1}{2n} \sum_{i=1}^{N} \sum_{\substack{j=1 \\ i\neq j}}^{N} (y_i - y_j)^2 \frac{n(n - 1)}{N(N - 1)}$$

$$= \frac{n - 1}{N - 1} \sum_{i=1}^{N} (y_i - \bar{Y})^2$$

Consequently, in view of (17.11),

$$E\left[\frac{N(N - n)}{n(n - 1)} \sum_{i\in s} (y_i - \bar{y})^2\right] = \frac{n - 1}{N - 1} \frac{N(N - n)}{n(n - 1)} \sum_{i=1}^{N} (y_i - \bar{Y})^2$$

$$= E(\hat{Y} - Y)^2$$

Q.E.D.

RATIO ESTIMATE AND SIMPLE RANDOM SAMPLING This strategy is characterized by $\pi_i = n/N$, x_i, $t_i = 1/x_i$, and $w_i = X/x$, where x is the sample total:

$$x = \sum_{i\in s} x_i \tag{23.3}$$

The ratio estimate $\check{Y}$ equals

$$\check{Y} = X \frac{\Sigma_{i\in s}\, y_i}{\Sigma_{i\in s}\, x_i} = X \frac{y}{x} \tag{23.4}$$

where y is also the sample total.

Solving

$$\sum_{i\in s} (y_i - \beta x_i)^2 \frac{1}{x_i} = \text{minimum} \tag{23.5}$$

we obtain

$$\hat{\beta} = \frac{y}{x} \tag{23.6}$$

and next

$$\hat{\beta} x_i \Big|_{y_i=1, y_j=0, j \neq i} = \frac{x_i}{x} \qquad 1 \le i \le N \tag{23.7}$$

Substituting (23.6) and (23.7) into (22.14), we obtain

$$\hat{E}(\check{Y} - Y)^2 = \sum_{i \in s} \left(y_i - \frac{y}{x} x_i \right)^2 \frac{X(X - x)}{x(x - x_i)} \tag{23.8}$$

If $x_i \equiv 1$, (23.8) coincides with (23.2), as it should be.

REGRESSION ESTIMATE AND SIMPLE RANDOM SAMPLING This strategy is characterized by $\pi_i = n/N$, $t_i = 1$, and two regressors, x and 1. The estimate is given by (19.24) and the sample estimates $\hat{\alpha}$ and $\hat{\beta}$ of the regression coefficients in (19.23). The weights w_i equal

$$w_i = \frac{N}{n} - \frac{(x_i - \bar{x})(N\bar{x} - X)}{\Sigma_{j \in s} (x_j - \bar{x})^2} \tag{23.9}$$

Now

$$\hat{\varepsilon}_i = \hat{\alpha} + \hat{\beta} x_i \Big|_{y_i=1, y_j=0, j \neq i} = \frac{1}{n} + \frac{(x_i - \bar{x})^2}{\Sigma_{j \in s} (x_j - \bar{x})^2} \tag{23.10}$$

Thus (22.14) yields

$$\hat{E}(\tilde{Y} - Y)^2 = \sum_{i \in s} [y_i - \bar{y} - \hat{\beta}(x_i - \bar{x})]^2 w_i (w_i - 1)(1 - \hat{\varepsilon}_i)^{-1} \tag{23.11}$$

where w_i and $\hat{\varepsilon}_i$ are given in (23.9) and (23.10).

If we wish to simplify further, we may put

$$\frac{\hat{v}_i}{\pi_i} \doteq w_i (w_i - 1) \frac{n}{n - 2} \doteq \frac{N(N - n)}{n(n - 2)} \tag{23.12}$$

Using the simplest expression for $\hat{v}_i/\pi_i$, we have

$$\hat{E}(\tilde{Y} - Y)^2 = \sum_{i \in s} [y_i - \bar{y} - \hat{\beta}(x_i - \bar{x})]^2 \frac{N(N - n)}{n(n - 2)} \tag{23.13}$$

SIMPLE ESTIMATE AND STRATIFIED SIMPLE RANDOM SAMPLING If the population and sample sizes in the hth stratum are N_h and n_h, respectively, the simple estimate equals

$$\hat{Y} = \sum_{h=1}^{L} \frac{N_h}{n_h} \sum_{i \in s_h} y_i = \sum_{h=1}^{L} N_h \bar{y}_h = \sum_{h=1}^{L} \hat{Y}_h \tag{23.14}$$

As usual, s_h denotes the sample from S_h--the hth stratum, $\bar{y}_h$ is the respective sample average, and $\hat{Y}_h$ is the simple estimate of the hth stratum total Y_h.

Introducing the strata indicators $n_{hi} = 1$ if $i \in S_h$ and $n_{hi} = 0$ otherwise, the strategy is representative to all of them, because of the sampling method.

Applying (23.2) to the hth stratum, we have

$$\hat{E}(\hat{Y}_h - Y_h)^2 = \frac{N_h(N_h - n_h)}{n_h(n_h - 1)} \sum_{i \in s_h} (y_i - \bar{y}_h)^2$$

Since the estimates $\hat{Y}_h$ are independent and therefore (20.5) holds, we are justified in putting

$$\hat{E}(\hat{Y} - Y)^2 = \sum_{h=1}^{L} \hat{E}(\hat{Y}_h - Y_h)^2$$

$$= \sum_{h=1}^{L} \frac{N_h(N_h - n_h)}{n_h(n_h - 1)} \sum_{i \in s_h} (y_i - \bar{y}_h)^2 \tag{23.15}$$

From Theorem 23.1 and from (20.5) it follows that (23.15) is an unbiased estimate of $E(\hat{Y} - Y)^2$.

Formula (23.15) is also a special case of the general formula (22.14) for $m = L$, $w_i = N_h/n_h$, $i \in s_h$, $\hat{\beta}_h = \bar{y}_h$, and $x_{hi} = n_{hi}$, $1 \le h \le L$.

RATIO ESTIMATE AND STRATIFIED SIMPLE RANDOM SAMPLING This strategy utilizes the estimate (20.19), which in our case may be rewritten as follows:

$$\check{Y} = X\frac{\hat{Y}}{\hat{X}} = X\frac{\sum_{h=1}^{L} N_h\bar{y}_h}{\sum_{h=1}^{L} N_h\bar{x}_h} \tag{23.16}$$

If the allocation is uniform, i.e., $n_h/N_h = n/N$, $1 \le h \le L$, then $\check{Y}$ simplifies to (23.4). The present strategy is representative with respect to variables of the form

$$y_i = \alpha_h + \beta x_i \qquad i \in S_h \qquad 1 \le h \le L \tag{23.17}$$

where the parameters α_h satisfy the side condition

$$\sum_{h=1}^{L} \alpha_h N_h = 0 \tag{23.18}$$

As may be seen from (20.20) for $z_i = 1$ the population estimates of α_h and β are

$$\bar{\beta} = \frac{Y}{X} \qquad \text{and} \qquad \bar{\alpha}_h = \bar{Y}_h - \bar{\beta}\bar{X}_h \qquad 1 \le h \le L \tag{23.19}$$

Corresponding sample estimates $\hat{\beta}$ and $\hat{\alpha}_h$ are obtained by *two-stage* least squares: We first get $\hat{\beta}$ by solving

$$\sum_{h=1}^{L} \sum_{i\in s_h} (y_i - \beta x_i)^2 \frac{N_h}{n_h} = \text{minimum} \tag{23.20}$$

and then $\hat{\alpha}_h$ by solving

$$\sum_{i\in s_h} (y_i - \hat{\beta}x_i - \alpha_h)^2 = \text{minimum} \qquad 1 \le h \le L \tag{23.21}$$

The solution of (23.20) is

$$\hat{\beta} = \frac{\sum_{h=1}^{L} N_h\bar{y}_h}{\sum_{h=1}^{L} N_h\bar{x}_h} = \frac{\hat{Y}}{\hat{X}} \tag{23.22}$$

and (23.21) yields

$$\hat{\alpha}_h = \bar{y}_h - \hat{\beta}\bar{x}_h \qquad 1 \le h \le L \tag{23.23}$$

It is easy to see that the $\hat{\alpha}_h$ satisfy the side condition (23.18).

In Theorem 23.1 we assumed that the $\hat{\beta}_k$ were obtained by one-stage least squares. However the concept of tightness and Theorem 23.1 may be extended to any estimation method such that

$$y_i - \sum_{k=1}^{m} \hat{\beta}_k x_{ki} = 0 \qquad i \in s \tag{23.24}$$

as soon as y_i coincides with any variable to which the strategy is representative. This is true in the present case, as may be easily checked. Consequently, we are justified in utilizing formula (22.14). Now, in view of (23.16),

$$w_i = \frac{N_h}{n_h}\frac{X}{\hat{X}} \qquad i \in s \tag{23.25}$$

and

$$\hat{\varepsilon}_i = \hat{\alpha}_h + \hat{\beta}x_i \Big|_{y_i=1, y_j=0, j\neq i} = \frac{1}{n_h} + \frac{N_h(x_i - \bar{x}_h)}{n_h\hat{X}}$$

$$\doteq \frac{1}{n_h} \qquad i \in s_h \qquad 1 \le h \le L \tag{23.26}$$

Substituting (23.22), (23.23), (23.25), and (23.26) into (22.14), we obtain

$$\hat{E}(\check{Y} - Y)^2 = \left(\frac{X}{\hat{X}}\right)^2 \sum_{h=1}^{L} \sum_{i\in s_h} \left[y_i - \bar{y}_h - \frac{\hat{Y}}{\hat{X}}(x_i - \bar{x}_h)\right]^2$$

$$\times \frac{N_h(N_h - n_h\hat{X}/X)}{n_h(n_h - 1)} \tag{23.27}$$

REFERENCES

Bryant, E. C., Hartley, H. O., and Jessen, R. J. (1960). Design and estimation in two-way classification. *J. Amer. Stat. Ass.* *55*, 105-124.

Cochran, W. G., Hartley, H. O., and Rao, J. N. K. (1962). On a simple procedure of unequal probability sampling without replacement. *J. Roy. Stat. Soc. Ser. B24*, 482-491.

Cramér, H. (1946). *Mathematical Methods of Statistics.* Princeton University Press, Princeton, N.J.

Dupačová, J. (1979). A note on rejective sampling. *Contributions to Statistics* (J. Hájek memorial volume), Academia Prague, 71-78.

Durbin, J. (1953). Some results when units are selected with unequal probabilities. *J. Roy. Stat. Soc., Ser. B 15*, 262-269.

Durbin, J. (1967). Estimation of sampling error in multistage sampling. *Appl. Stat. 16*, 152-164.

Ericson, W. A. (1969). Subjective Bayesian models in sampling finite populations. *J. Roy. Stat. Soc., Ser. B 31*, 195-233.

Fellegi, I. P. (1963). Sampling with varying probabilities without replacement: Rotating and non-rotating samples. *J. Amer. Stat. Ass. 58*, 183-201.

Godambe, V. P. (1955). A unified theory of sampling from finite populations. *J. Roy. Stat. Soc., Ser. B 17*, 269-278.

Godambe, V. P., and Joshi, V. M. (1965). Admissibility and Bayes estimation in sampling finite populations I, II. *Ann. Math. Stat. 36*, 1706-1722, 1723-1742.

Goodman, R., and Kish, L. (1950). Controlled selection: A technique in probability sampling. *J. Amer. Stat. Ass. 45*, 350-372.

Hájek, J. (1949). Representativní výběr skupin metodou dvou fází (Representative sampling by the method of two phases, in Czech). *Statisticky Obzor 29*, 384-395.

Hájek, J. (1959). Optimum strategy and other problems in probability sampling. *Casopis Pro Pěstování Matematiky* (Journal for Cultivation of Mathematics) *84*, 387-423.

Hájek, J. (1964). Asymptotic theory of rejective sampling with varying probabilities from a finite population. *Ann. Math. Stat. 35*, 1491-1523.

Hald, A. (1952). *Statistical Theory with Engineering Applications.* Wiley, New York and London.

Hanurav, T. V. (1967). Optimum utilization of auxiliary information: π_{ps} sampling of two units from a stratum. *J. Roy. Stat. Soc., Ser. B 29*, 374-391.

Hartley, H. O., and Rao, J. N. K. (1962). Sampling with unequal probabilities and without replacement. *Ann. Math. Stat. 33*, 350-374.

Horwitz, D. G., and Thompson, D. J. (1952). A generalization of sampling without replacement from a finite universe. *J. Amer. Stat. Ass. 47*, 663-685.

Joshi, V. M. (1965). Admissibility and Bayes estimation on sampling finite populations, III. *Ann. Math. Stat. 36*, 1730-1742.

Joshi, V. M. (1969). Admissibility of estimates of the mean of a finite population, in *New Developments in Survey Sampling* (N. L. Johnson and H. Smith, Jr., eds.). Wiley-Interscience, New York and London.

Lahiri, O. B. (1951). A method for selection providing unbiased estimates. *Int. Stat. Ass. Bull. 33*, 133-140.

Madow, W. G. (1949). On the theory of systematic sampling, II. *Ann. Math. Stat. 20*, 333-354.

Midzuno, H. (1951). On the sampling system with probabilities proportionate to sum of signs. *Ann. Inst. Stat. Math. 2*, 99-108.

Narain, R. D. (1951). On sampling without replacement with varying probabilities. *Ind. Soc. Agric. Stat. 3*, 169-174.

Petrov, V. V. (1972). *Sums of Independent Random Variables.* Springer-Verlag, Berlin.

Raj, D. (1956). A note on the determination of optimum probabilities in sampling without replacement. *Sankhyā 17*, 197-200.

Rao, C. R. (1965). *Linear Statistical Inference and Its Applications.* Wiley, New York and London.

Rao, J. N. K. (1963). On three procedures of unequal probability sampling without replacement. *J. Amer. Stat. Ass. 58*, 202-215.

Rao, J. N. K. (1967). On the choice of a strategy for a ratio method estimation. *J. Roy. Stat. Soc., Ser. B 29*, 392-397.

Rosén, B. (1970). On the coupon collector's waiting time. *Ann. Math. Stat. 41*, 1952-1969.

Rosén, B. (1972). Asymptotic theory for successive sampling with varying probabilities without replacement, I, II. *Ann. Math. Stat. 43*, 373-397, 748-776.

Sampford, M. R. (1967). On sampling without replacement with unequal probabilities of selection. *Biometrika 54*, 499-513.

Sampford, M. R. (1969). A comparison of some possible methods of sampling from smallish populations, with units of unequal size, in *New Developments in Survey Sampling* (N. L. Johnson and H. Smith, Jr., eds.). Symposium held at Chapel Hill, N. Car. Wiley-Interscience, New York and London.

Samuels, S. M. (1965). On the number of successes in independent trials. *Ann. Math. Stat. 36*, 1272-1278.

Yates, F., and Grundy, P. M. (1953). Selection without replacement from within strata with probability proportionate to size. *J. Roy. Stat. Soc., Ser. B 15*, 236-261.

INDEX